Palgrave Innovations in Philosophy

Palgrave Innovations in Philosophy is a new series of monographs. Each book in the series will constitute the 'new wave' of philosophy, both in terms of its topic and the research profile of the author. The books will be concerned with exciting new research topics of particular contemporary interest, and will include topics at the intersection of Philosophy and other research areas. They will be written by up-and-coming young philosophers who have already established a strong research profile and who are clearly going to be leading researchers of the future. Each monograph in this series will provide an overview of the research area in question while at the same time significantly advancing the debate on this topic and giving the reader a sense of where this debate might be heading next. The books in the series would be of interest to researchers and advanced students within philosophy and its neighboring scientific environments.

More information about this series at
http://www.palgrave.com/gp/series/14689

Nenad Miščević

Curiosity as an Epistemic Virtue

palgrave
macmillan

Nenad Miščević
Philosophy
University of Maribor
Maribor, Slovenia

Palgrave Innovations in Philosophy
ISBN 978-3-030-57102-3 ISBN 978-3-030-57103-0 (eBook)
https://doi.org/10.1007/978-3-030-57103-0

This Palgrave Macmillan imprint is published by the registered company Springer Nature Switzerland AG.
The registered company address is: Gewerbestrasse 11, 6330 Cham, Switzerland

Preface

Many colleagues and friends have contributed to the genesis of the book. My work on curiosity has found the initial impetus in reading Ilhan Inan's book on the topic, and then enjoying conversations with him—official in conferences, and unofficial in nice places in Dubrovnik, Istanbul and the like.

Next, I wish to thank Ernest Sosa, whose work on virtue epistemology has been influencing me for two decades; we had nice discussion of curiosity in places like Taiwan and Bled in Slovenia. Duncan Pritchard was a precious conversation partner, whose influence is visible in particular in connection with problems of epistemic value of curiosity.

Thanks also go to J. Baehr, M. Slote, L. Watson and M. Watkins, with whom I have discussed curiosity in Rijeka, 2014.

Nenad Miščević
Maribor, Slovenia

Contents

List of Figures

Part I

Understanding Curiosity

1

Introduction

Introduction

The desire to know, or curiosity or inquisitiveness, has been for more than two millennia discussed in philosophical literature, under various, not completely synonymous, names. Curiosity is thus an old topic in classical philosophy; however, it is a new area of research in contemporary epistemology. It had almost disappeared from the twentieth-century philosophical scene, in particular from analytic debates, which have concentrated on the definition of knowledge rather than on its goals and its motivating sources. Fortunately, it is back: in the recent virtue epistemology there is a kind of upsurge of interest in it. On the descriptive epistemological side Ilhan Inan has published a book (2012) on the semantics of curiosity. There is an excellent recent (2018) edited volume on the moral psychology of curiosity, as its title goes. The present book will probably be the first book on curiosity from the strictly epistemological viewpoint, but also taking into account the recent developments in psychology of curiosity and in the research on the nature and motivation of inquiry in science.

N. Miščević, *Curiosity as an Epistemic Virtue*, Palgrave Innovations in Philosophy,
https://doi.org/10.1007/978-3-030-57103-0_1

The viewpoint taken here will be broadly the one of virtue epistemology (VE for short). In short, in the present book we want to bring curiosity where it belongs, according to our opinion, namely in the very center of epistemology. We shall treat is as the central epistemic virtue, and we shall say more about it in a moment.

The chapter you are reading is a topical introduction; the next chapter will offer a sketchy historical introduction (sketchy and short because of the demands of space, with apologies).

Let us start with terminology. We shall be using the word "curiosity" as our central term; sometimes we shall also use "inquisitiveness", to stress its active side.[1] We shall take it to be the interest in how things are, theoretically and practically, the desire for knowledge and understanding (for this meaning see, for instance, Baehr 2011); this will be the main curiosity concept to be used here. Sometimes the word "curiosity" is used in a negative sense, of meddlesomeness; this is not the sense to be used here. (For more on this, see Chap. 4.)

As we noted, not much has been written on epistemology of curiosity recently. The excellent monograph written by Inan (2012) combines semantic and epistemological approach, with a bit more stress being laid on semantics. The (2018) collection we mentioned offers an impressive range of approaches, from some of the best philosophers in the field; we shall be referring to both books in the sequel.

There is more to be found in the related areas. First, the history of curiosity and of approaches to it is quite rich; the ancient and early Christian philosophy is discussed in detail (e.g. Zuss 2012; Walsh 1988; Zurn 2019), and the further development has been masterly analyzed by Hans Blumenberg (1988; the English version appears as Part III, "The 'Trial' of Theoretical Curiosity", of Hans Blumenberg (1966), *The Legitimacy of the Modern Age*, MIT Press). Neil Kenny (2004) offers a detailed account of the "rapprochement" between the curious and the useful in early modern times. There are also a lot of interesting studies about the role of curiosity in science, and in particular in scientific revolution (see Chap. 9 on science). Second, equally importantly, a lot of

[1] With thanks to Safiye Yiğit, who has, in the discussion, insisted on the importance of the term (see also her 2018 paper).

work on curiosity has been done in cognitive psychology, by authors like Berlyne, Silvia and Engel; see Chap. 8 on cognitive psychology of curiosity.

Understanding Curiosity

What is curiosity? Let us start with delineating a conception of curiosity in the most general sense of motivation for acquiring knowledge and understanding, and by noting some important distinctions, often overlooked by friends and foes of curiosity alike. One usually distinguishes between the related disposition, sometimes called "interest", especially by psychologists (see Silvia 2006), and the manifestation of the disposition (more like inquisitiveness, or active curiosity, perhaps), going from the less known to the better known (see Inan 2012). In fact, one should distinguish between the general interest in things, the capacity and readiness cognitively to react to features of the environment, curiosity proper and the disposition to ask questions and inquire. Psychologists are interested in the issue whether curiosity is mere desire or emotion; the latter view seems dominant (see Silvia 2006). However, here the simpler, desire-like aspect will be sufficient. When I have active curiosity in mind, I shall sometimes talk of "inquisitiveness-curiosity", to remind the reader that we are dealing with curiosity in one of its varieties, not with some other, related phenomenon.

"All man by nature desire to know", Aristotle famously claimed in *Metaphysics, A. 1* (Ross' translation), and, of course, the core element in knowledge is true belief. Why do we want to have true beliefs about very diverse matters that interest us in life? Because we are curious about things, inquisitive and alert, and inquisitiveness-curiosity regarding *p* is the wish to have true beliefs and to know whether *p* (and to understand why *p*, etc.). This sounds quite banal and uncontroversial to many. However, Ernest Sosa, who otherwise keeps stressing the importance of truth in epistemology, has argued that the wish to have answers to questions we are curious about cannot be put in terms of "desire for truths per se" (2002, 158). He takes the desire-for-truths theorist to make a fallacy, which he illustrates by the analogy with the desire for savory food. He invites us to imagine a character claiming:

P1 I want savory food.

and

P2 I want that *if* I have savory food, it be also nutritious.

and then concluding from this:

C Therefore, I want nutritious food.

The desire-for-truths theorist allegedly makes the same mistake by arguing:

F1 I want beliefs that answer my questions.
F2 I want that *if* I have an answer to a question of mine, it be true!
C Therefore, I want true beliefs. (2002: 158)

Next, Sosa admonishes us:

> We may want true beliefs, in this sense: that *if* for whatever reason. we are interested in a certain question. we would prefer to believe a correct rather than an incorrect answer to that question: but this does not mean that we want, in itself and independently of our wanting our questions answered, that we have true answers to them simply for the truth this would give us. (Ibid.)

For my part, this is not how I see my own curiosity. It is not that when I ask you, say for time, I want *an* answer and then, in addition, I want the true one, like wanting savory food and in addition wanting it to be nutritious. I don't want you just to say "It's five p.m." and then have an additional wish that your sentence come out true; I want it to be true in the first place. A sign that I am not being idiosyncratic is that in the movies, police investigators, when given an answer by the suspect, never say "O thanks that you answered, but it would be nice if your answer were also true"; they typically shout "I want the truth!" I can't believe that they are all into committing the desire-for-truths theorist's alleged mistake. I conclude that there is nothing abnormal about desiring only true answers if one desires any answers at all. Inquisitiveness in general is the disposition to have such desires and wishes and to pursue their fulfillment.

So, why are we inquisitive? Our inquisitiveness-curiosity is either pure or practical or mixed. Sometimes, one is just curious, with no further practical goal. This can be called "intrinsic curiosity". But one is often motivated extrinsically, by practical curiosity, searching the means for practical ends. Classics did think about the issue. La Rochefoucauld distinguishes two kinds of practical goals:

> There are various sorts of curiosity; one is from interest, which makes us desire to know that which may be useful to us; and the other, from pride which comes from the wish to know what others are ignorant of. (*Maxims*)

Here we shall concentrate upon the pure variety. A human being devoid of curiosity would have little motivation to arrive to true belief and knowledge.[2] In normal cases it is inquisitiveness-curiosity that motivates us to gain true belief and knowledge. On the usual view of motivating virtues, this would seem to make it a virtue; since it is the main spring of motivation, we should take it as *the motivating epistemic virtue. After all, wanting to know whether p* it gives cognizers particular instances of *p* (or of its negation) as particular goals and the truth as the general epistemic goal. So, we have a truth-focused motivating virtue: *inquisitiveness or curiosity* having as its general goal reliable arriving at truth. This is, I submit, the *core motivating epistemic virtue.* There is a multitude of questions of all sorts that we ask, whether, why, when and how, and inquisitiveness-curiosity caters to all of them.

I have just mentioned being curious and alert. I propose to take the notion of inquisitiveness-curiosity in the widest sense so as to encompass

[2] We are here considering the cognizer in isolation from social structures of inquiry. If I am part of a research institution, I might, „be motivated at arriving at true belief, because otherwise I will be fired", if I am a private eye, I might have to investigate other people's marital infidelities that I find personally very boring and uninterested. Here the curiosity is simply institutionalized.

primitive alertness to the features of the environment. (If this doesn't fit your semantic intuitions, then take it, please, as a stipulation, not as analysis of the commonsense notion). Alertness is biologically based, as shown by the existence and functioning of early warning perceptual mechanisms that alert the organism to potentially threatening changes in its surrounding, by the mechanism of habituation, that makes it "lose interest" in repeated stimuli. The importance of this wide sense of inquisitiveness is that it helps us address the problem of "brute" or "passive" knowledge, as Jason Baehr calls it (Baehr 2011). The answer is to take alertness as proto-curiosity, and in this sense, a proto-virtue that makes us open to the world and sensitive to it.[3]

We need one more widening. As various colleagues have noticed, a person, finite or infinite, who knows everything would not be curious, and would thus paradoxically lack the alleged main motivating epistemic virtue. One answer is that many human virtues are tailor-made for human agents in less-than-perfect but more-than-hellish human circumstances. We shall argue in Chap. 3 that curiosity is one such virtue, typical for finite and relatively ignorant beings, in need of constant updating of information in order to function successfully. But I prefer another line: I will just stipulate a slightly wider meaning of "inquisitiveness" that also includes cherishing the truth once found. It seems to me a natural extension of meaning: a person with bad memory but eager to get to know who subsequently doesn't care a bit for the knowledge acquired and is completely unworried about having forgotten everything she learned is not consistently inquisitive. So, the hypothetical omniscient person who keeps her virtue by cherishing what she knows is "curious" in this wider sense.

And finally, a slight narrowing of the goal. As my Bulgarian colleague Bakalova has reminded me, curiosity sometimes leads to insights, and some of the insights are not directed toward truth like hitting upon a great idea in poetry, or choreography. Of course the proposed account has no problem with curiosity sometimes aiming at items other than

[3] Alertness is beyond our control, one might object. Well, in many cases curiosity is beyond our control as well, and being under one's control is not essential for motivating virtue. But I agree that the automatism with which we do get alert makes alertness closer to the "sub-personal level" than inquisitiveness.

truth; it only claims that in epistemically relevant cases it does aim at truth. Note that there is a link with knowledge even in this case, since the insight sought would typically have to do with knowledge-how.[4] So much, or rather, and unfortunately, so little, about the very notion.[5]

The Preview of the Book

Overview

Since the very beginning of the book we put curiosity at the center of philosophical interest, combining the initial sketch of kinds of curiosity with deeper epistemological issues related to them. The book has three parts: a more introductory one (preparing the ground for the second, central one); the central part on curiosity as motivating and organizing virtue, possibly the central epistemic virtue tout court; and the third, dedicated to ramifications and applications. The central part also tentatively addresses the issue of epistemic value, proposing that a large part of it derives from our natural curiosity. The third part talks about cognitive psychology of curiosity, about curiosity in science and, finally, about Socratic self-inquisitiveness or self-curiosity and its possible role in wisdom. Let me pass through the main topics in order.

History: How It All Started

The historical chapter is a kind of "historical introduction" to the topic. It presents not the whole of history, since this would demand a book, but the beginnings of philosophical discussion of curiosity, noting how different evaluations of curiosity started and starting a story of good and bad times for curiosity as philosophical topic. So, it gives on overview of

[4] One might go even further in discussion and claim, with Stanley and Williamson (2001), that knowledge-how is a subspecies of knowledge-that, but I am not particularly enthusiastic about this line.

[5] Horwich (2006) discusses similar issues under the heading of truth goal; I find his remarks congenial, but it is remarkable that he never mentions inquisitiveness, nor the topic of intellectual virtues.

ancient admirers and critics of curiosity from Greek atomists, through Socrates and Plato, all the way to Stoics and to early Christian thinkers, culminating with the rich and refined account proposed by Saint Augustine. It very briefly turns to the early modern "revolution of curiosity", within general philosophy and in areas in which philosophy has been joining early modern science, with names such as Galileo, Descartes, Hume and Kant at the forefront.

Kinds of Curiosity

The ancient discussion has already offered a rich taxonomy of kinds of curiosity, often organized around the positive-negative contrast. Modern work has added more, and also changed the focus from value-centered divisions to more topical ones.

Curiosity as Virtue

A human being devoid of curiosity would have little motivation to arrive to true belief and knowledge. Scientists, from Darwin through Einstein to Hawking, have spoken about curiosity as their crucial motivation. On the usual view of motivating virtues, this would seem to make it a virtue; since it is the main spring of motivation, we should take it as the motivating epistemic virtue. Indeed, philosophers have traditionally recommended live interest in at least certain important areas: one should come to know oneself, one should study important features of nature, or of supernatural reality.

However, many philosophers thought curiosity is not a virtue. Plutarch describes it in completely negative terms, as an "unhealthy and injurious" state of mind which allows "winter and darkness to enter the soul". There are problems with bad curiosity: internal, epistemic problems (the temptation to study marginal, unimportant matters, stressed by Brady and Sosa) and the moral ones (interest in base matters, abuse of scientific curiosity for helping aggressive, cruel and otherwise immoral behavior). Can curiosity still be considered a virtue?

The temptation to study of superficial matters seems to be well documented in cognitive psychology, as Michael Brady has recently argued. I defend curiosity against the accusation and argue that he has proposed a needlessly pessimistic reading of the cognitive data. The morally bad aspects of curiosity cannot be denied. I argue that other mainstream virtues like courage (among the moral ones) and epistemic modesty (among the epistemic ones) are plagued by analogous problems. The range of possible solutions is well known; I propose as the two acceptable ones either limiting the virtue status to "good" curiosity (as it is often done with courage and the like, e.g., by Foot (2002)) or building into virtuous curiosity the ability to recognize the right objects to deal with, and admissible situations for the exercise of curiosity.

Why Is Curiosity Central: A Motivating Virtue Account?

It is epistemically very important to be *intrinsically* motivated to acquire knowledge and understanding. However, most of the character virtues apart from curiosity do not motivate such acquisition. Of course, virtues-abilities are not motivating in themselves; they help realize the goals we are independently motivated to achieve. The genuine curiosity is the central intrinsically motivating drive for achieving knowledge and understanding. We shall see an example in Chap. 5, the role of self-inquisitiveness in organizing the acquisition of self-knowledge. Curiosity is the core motivating epistemic virtue. There is a multitude of questions of all sorts that we ask, whether, why, when and how, and inquisitiveness-curiosity caters to all of them.

So, in the present chapter, a strong, strictly virtue-*based*, and at the same time truth-centered framework for virtue epistemology (VE) is proposed, It bases VE upon a clearly motivating epistemic virtue, inquisitiveness or curiosity in a very wide sense, characterizes the purely executive capacities-virtues as means for the truth-goal set by the former and, finally, situates the remaining, partly motivating and partly executive virtues in relation to this central stock of virtues. Character-traits epistemic virtues are presented as hybrids, partly moral, partly purely epistemic. In order to make the approach virtue *based*, it is argued that the central virtue, inquisitiveness or curiosity, is responsible for the value of truth: truth is valuable to cognizers because they are inquisitive, and most other virtues are means for satisfying inquisitiveness. One can usefully combine this virtue-based account of the motivation for acquiring knowledge with an analysis of the concept "knowledge", which puts at the forefront virtues-capacities, in order to obtain a full-blooded, "strong" VE. We call the result "the motivating virtue account".

The Motivating Virtue Account: For and Against

The high valuation of curiosity has prompted strong contemporary criticisms, which focus upon possible deficiencies of curiosity. First, upon the cases of idle curiosity, and the second, upon volatility, irrelevance, superficiality and similar defects. Finally, there is an objection that targets the idea of epistemic high usefulness of curiosity: excellent epistemic results can be obtained without intrinsic, paradigmatic curiosity, so what is epistemically so special about it? The chapter offers the defense of the motivating virtue account against these.

Curiosity and the Origin of Epistemic Value

A separate chapter addresses the deep metaphysical problem of the origin of epistemic value. Instrumental value raises no problem: the practical goals give value to items of knowledge that serve as means for them. But what about intrinsic value? Does curiosity bestow such value upon truths and knowledge of them? Or, the other way around, the value of truth-cum-knowledge determines the value of curiosity? The chapter offers an overview of positions, with authors like Goldman, Zagzebski and Pritchard on the front-line. My own sympathies are response-dependentist: the genuine intrinsic curiosity bestows intrinsic value on knowledge and truth, and I propose the view as a tentative solution. The proposed account is thus response-dependentist and naturalist.

Cognitive Psychology

We next turn to interdisciplinary matters. First, to the contribution of cognitive psychology to the characterization of curiosity and to one author, Paul Silvia, and his book *Exploring the Psychology of Interest* (2006). We show how the cognitive research goes well with philosophical interest, and how the role of understanding, central in the former, gets reaffirmed in the latter. We conclude with a lot of optimism about descriptive naturalization of virtue-epistemological approach to curiosity, which would, as against pure concept-analyzers, include natural, psychological conditions of possibility of epistemically virtuous activities.

Scientific Curiosity

Curiosity is the motivating force in science, we claim, engaged in deploying, focusing and helping organize our knowledge-capacities. A good classification of kinds of curiosity might turn out to be quite important for understanding scientific research: compare theoretical curiosity that leads to understanding and practical curiosity directly motivating laboratory work. The chapter investigates the ways the two interact, and the link between the desire to understand and the desire to learn how to apply the understanding reached.

We also briefly look at the cognitive structure of scientific revolutions, since it finely illustrates the roles that different kinds of curiosity have actually played in historical development.

Know Thyself: The Importance of Self-inquisitiveness for Wisdom

We next turn to the role of curiosity as epistemic virtue, taking into account the traditionally central kind of interest, namely the interest in knowing oneself and in examined life. Authors writing about the meaning of life, like Nozick, have continued the tradition, but virtue epistemologists have stayed away from it; I suggest that we take the "know thyself" drive as an important example of epistemic imperative, and turn to the topic from the standpoint of mainstream virtue epistemology.

Of course, self-inquisitiveness is a motivating epistemic virtue. If successful, it deploys, organizes and motivates other epistemic virtues, both virtues-abilities and character virtues. Self-inquisitiveness thus points to a possible general paradigm of curiosity organizing. And probably the intrinsic self-inquisitiveness is also responsible for the intrinsic value of self-knowledge.

Toward a Bigger Picture

We hope that in the book we succeed in bringing together two lines of inquiry. First, the virtue epistemological one, with curiosity as the central item, and second, the issue of the moral worth of curiosity, with the problems about bad curiosity and with curiosity about oneself, and its role in a meaningful life, as a paradigmatic positive case.

In the concluding chapter we list three further topics, worthy of investigating, which we leave for some later occasion.

First, philosophical curiosity; a topic that is sometimes mentioned, but rarely developed in a more systematic fashion. We hope to apply our ideas about the desire to understand, guiding the work of science, and to connect it to the work in philosophy. So the issue of what is central for philosophical curiosity, the desire for philosophical understanding, will be briefly sketched.

Second, the social framework of normal exercises of curiosity. There, the division of labor, in science, law and media, increases efficiency and creates now possibilities, but it also adds to existing inequalities and creates new ones, from economical through legal to political ones. The topic of epistemic justice in relation to inquiry and inquisitiveness is therefore a burning topic, and we briefly point out its importance for social epistemology. Also, we remind the reader about problems in social epistemology and philosophy of science having to do with social organization of curiosity, and its theoretical and moral consequences.

Third, we return to the topic of naturalism and widen our sketchy presentation from various previous chapters in the book.

We thus hope to show again that curiosity merits to become a very hot topic in epistemology and ethics, as it has been historically for centuries.

References

Baehr, Jason. 2011. *The Inquiring Mind: On Intellectual Virtues and Virtue, Epistemology*. Oxford: Oxford University Press.

Blumenberg, Hans. 1966. *The Legitimacy of the Modern Age*. MIT Press.

———. 1988. *Der Prozeß der theoretischen Neugierde*. Suhrkamp.

Foot. (2002). Virtues and Vices and Other Essays in Moral Philosophy, second edition (first edition 1978), Oxford: Oxford University Press, https://doi.org/10.1093/0199252866.001.0001

Horwich. (2006). Value of truth. *Noûs* 40 (2): 347–360.

Inan, Ilhan. 2012. *The Philosophy of Curiosity*. Routledge.

Kenny, Neil. 2004. *The Uses of Curiosity in Early Modern France and Germany*. Oxford University Press.

Silvia, Paul J. 2006. *Exploring the Psychology of Interest*. Oxford University Press.

Sosa. (2002). Tracking, competence and knowledge in Paul K. Moser (ed.), The Oxford Handbook of Epistemology. Oxford University Press. pp. 264–287.

Stanley, Jason, and Timothy Williamson. 2001. Knowing How. *The Journal of Philosophy* 98 (8): 411–444.

Walsh, Peter. 1988. The Rights and Wrongs of Curiosity (Plutarch to Augustine). *Greece & Rome* 35 (1): 73–85.

Yiğit, Safiye. 2018. Curiosity as an Intellectual Virtue. In *The Moral Psychology of Curiosity*, ed. Ilhan Inan, Lani Watson, Dennis Whitcomb, and Safiye Yiğit, 117–140. Rowman & Littlefield International Ltd.

Zurn, Peter. 2019. Busybody, Hunter, Dancer: Three Historico-Philosophical Models of Curiosity. In *Curious about Curiosity: Toward New Philosophical Explorations of the Desire to Know*, ed. M. Papastefanou, 26–49. Cambridge Scholars Press.

Zuss, Mark. 2012. *The Practice of Theoretical Curiosity*. Springer.

2

History: How It All Started

We now pass to a brief historical introduction to the problematics—not the whole of history, since that would demand a book, but a historical overview of the beginnings of philosophical discussion of curiosity, ending with a brief note on the apotheosis of curiosity in early modern philosophy. Our main interest will be in how philosophy of curiosity or "love of knowledge" started, and in particular, how different evaluations of curiosity started, since this connects to the central topic of the book—the issue of the virtuous versus vicious nature of curiosity.[1]

I shall be very brief about the most famous authors whose contributions are well known, and present at more length the views of Seneca,

[1] There are a lot of detailed studies, but little general literature on the question; the most important exceptions are the classical work by Blumenberg, Part III "The 'Trial' of Theoretical Curiosity", in Hans Blumenberg (1966), *The Legitimacy of the Modern Age*, MIT Press (our quotations will be from the 1999 edition) and then the relatively recent book by Mark Zuss (2012), *The Practice of Theoretical Curiosity*, Springer. (See also the review and discussion by Gene Fellner, Wesley Pitts, and Mark Zuss (2012), "Beyond the Sensible World: A Discussion of Mark Zuss", *The Practice of Theoretical Curiosity*, *Cult Stud of Sci Educ*, online publication without page numbers.)

For a recent discussion, see the excellent paper by Safiye Yigit (2019), "The Curious Case of Curiosity: A virtue or a Vice?" in Marianna Papastephanou (ed.), *Toward New Philosophical Explorations of the Epistemic Desire to Know: Just Curious about Curiosity*, Cambridge Scholars Publishing, 150–165.

N. Miščević, *Curiosity as an Epistemic Virtue*, Palgrave Innovations in Philosophy,
https://doi.org/10.1007/978-3-030-57103-0_2

Cicero and Plutarch, who are being less discussed in the general literature. Apologies for the extreme brevity: the usual approach is to separate the topical from the historical approach and then give a lot of space to each, a separate book, normally. However, with the topic of curiosity, the history has not been discussed in the topical analytic literature, so I think that even a very short introduction might be very useful to the contemporary reader!

The Classical Period

Let us start with a few words about the Pre-Socratics. With them the search for knowledge becomes a central aim of human endeavors. The alleged vast knowledge of gods is admired, and the question of comparison with human abilities raised. The knowledge in question concerns cosmos, but also humans.

Let me illustrate this by reference to Xenophanes of Colophon, who was a philosophically minded poet who lived in various parts of the ancient Greek world during the late sixth and early fifth centuries BCE. Xenophanes' most extended comment on knowledge is B34:

> [A]nd of course the clear and certain truth no man has seen nor will there be anyone who knows about the gods and what I say about all things. For even if, in the best case, one happened to speak just of what has been brought to pass, still he himself would not know. But opinion is allotted to all.

The commenters (e.g. Zuss 2012: 11) note that Xenophanes claimed that our knowledge will always remain incomplete and obscure. While wisdom remains always the ultimate goal of human aspiration, our ability to know the ways of the world are only developed by constant search. Zuss ascribes to Xenophanes a positive attitude to rational inquiry:

> He asserts the right to rational inquiry through observation, debate, and a discourse that, perhaps for the first time in the classical context, bridges the gulf between the gods and humanity. (Ibid.)

Other thinkers, like Hecateus, join the positive project (Zuss 2012). We should equally not forget the genuinely curious philosopher Leucippus, from fifth century BCE, who allegedly said that "he would rather discover a single explanation (aitiologian), than acquire the kingdom of the Persians" (from Eusebius). So, much about Pre-Socratics, again with apologies for brevity.

Now, with Socrates a contrast appears that will become canonical for a large part of ancient philosophy: the one between an interest in nature and cosmos, and an interest in human matters, including interest in one's own character, goals and so on, the one that we called self-inquisitiveness. Indeed, self-examination plays for him the central role in human life: "The unexamined life is not worth living", he famously claimed in the *Apology* (38a5–6).[2] The self-examination goes together with ethical inquiry.[3]

Here is Richard Kraut commenting on Socrates:

> [H]e holds that ethical inquiry is a process that one should undertake throughout one's life, not merely for some brief period. One cannot live up to his demand by spending a half year asking the questions he asks, then turning to other matters, and never revisiting such issues. For the call to the examined life is linked to the thesis that the greatest good for a human being is "to have discussions every day about virtue" and other ethical matters (*Apology* 38a3). We should recognize how audacious a claim this is. We should expect Socrates to give us reasons to accept it. (2006: 229)

And Kraut points to the way from Socrates to Plato. He notes a point common to the major moral philosophers of antiquity:

> Above all, they think, one must arrive at an understanding, far beyond that of a child, of what is good. That is the principal concept of Greek ethics, and the Socratic dialogues lay the groundwork for its centrality [...] The

[2] See, for example, the comments of Christopher Rowe in his 2011 study of self-examination in *Apology*.

[3] For the contemporary perspective on self-examination see Chap. 10 of the present book.

> highest kind of knowledge, Plato holds in the *Republic*, is knowledge of the Form of the good—and it takes many years of scientific training to acquire it. (2006: 241)

Plato has indeed managed to bring together the two main lines of inquiry, and, in our terminology, two main targets of curiosity. On the one hand, he continues the Socratic line of importance of examination of humanly relevant matters, including self-examination and search for principles of good human life; on the other, he combines it ingeniously with the second line, inquiry into cosmos and its transcendent ground. He adds new elements to both lines, prominently the connection of inquiry with love (*agape*) for its target, including considerations of Eros and the like (see, for instance, *Phaedrus*, 256).

In the *Republic* Plato talks about "the real lover of knowledge" (*philomathes*), ready "to strive emulously for true being" (*pros to on*) ([490a]).[4] The discussion of the desire for knowledge thus gets combined with metaphysical considerations of its target. The final target cannot be the multiplicity of individuals but must be the true nature (*physis*) of each thing ([490b 3]).

Interestingly, the historians of the notion of curiosity, like, prominently Zuss, seem to be at pains with the terminology Plato uses: his "love of knowledge" does not fit all the connotations of the word "curiosity", so Zuss and others seem not to appreciate properly the impressive task of unification of the two lines of interest, the one in human and the other in cosmic-metaphysical matters, achieved by him.

Aristotle continues in building up the happy synthesis, of course with less romantic connections than his teacher. We have already quoted his famous line from *Metaphysics* I, to the effect that all human beings by nature desire to know. Here is the wider context:

[4] Here is the reading due to Martha Nussbaum:

> *The Republic* argues that the best life for a human being is the life of the philosopher, a life devoted to learning and the contemplation of truth.

> All men naturally desire knowledge (*tou eidenai oregontai physei*). An indication of this is our esteem for the senses; for apart from their use we esteem them for their own sake, and most of all the sense of sight. Not only with a view to action, but even when no action is contemplated, we prefer sight, generally speaking, to all the other senses. The reason of this is that of all the senses sight best helps us to know things, and reveals many distinctions. ([980a] [21])

Note the terminological richness of characterizations the drive for knowledge and truth in both philosophers. Here it is *orexis*, in Plato it was *agape* and also *philomathia,* and other names could also be found. Let me also mention the connection between desire to know and wonder. The latter is a manifestation of the former, and often its first phase (*Met.* 980a21); Aristotle here continues considerations on wonder started by Plato in *Theaetetus* where he notes that wonder is the "beginning of philosophy" (155d2–5). Now, in *Metaphysics* and in *Posterior Analytics* it is the metaphysical line that is prominent; however, an analogous approach is present in his works on ethics. Let me quote Kraut again:

> The student of ethics, Aristotle says, is embarked on the project of trying to become a better person, and in order to do so, he must come to a better understanding of the chief good of human life. Like an archer aiming at a target, he will be better able to hit his mark—living and making choices as he should—after having come to see, through philosophical argument, what his mark really is (*Nicomachean Ethics* I.2). (241)

Let me note a point that will become important later in the book, namely the fact that the meaning of Aristotle's word for "knowing (*epistasthai*) comes close to "understanding". We "know" something in this sense, he writes in Chapter 2 of his *Posterior Analytic* (71 b 10–18) when we have an explanation of it, and when we are aware that things cannot be otherwise.[5] For Aristotle, this "understanding" is the highest kind of knowledge, and it is natural to assume that the natural desire of human beings for knowledge also, and very importantly, encompasses the desire to understand. We shall return to this point when talking about kinds of

[5] The translation is from J. Barnes (1975). See also M. F. Burnyeat (1981, 2011).

curiosity and their psychology and about the desire to understand in science (Chap. 9).

To conclude, with Plato and Aristotle curiosity, the desire for truth, knowledge and understanding, under its various names, becomes recognized as a general driving force behind human effort and the center of human cognitive life.

Hellenistic Thought

Let us now pass to the Hellenistic period. Here, several crucial innovations take place. After Plato and Aristotle, the authors turn to more human concerns, and to the role of curiosity in daily life. They distinguish in detail various kinds of curiosity, use specific names for some of them, and evaluate the kinds in a strong moral manner. The typical contrasts are high (good) versus low (bad) curiosity, virtue-directed versus virtue-indifferent curiosity and the like. It is really a curiosity for-and-against debate, sometimes marked with extreme attitudes in either direction.

We shall follow chronological order, starting with Pyrrho (365/60–275/70 BC), passing to Cicero, (106–43 BC), Seneca (BC–65 CE), and then to Plutarch (45–120 CE), and conclude in the next section with Saint Augustine (354–430 CE). Our authors don't insist on metaphysical grounding of the desire for knowledge and are, for this reason, closer to the contemporary, purely epistemological and ethical approaches to curiosity. Interestingly, we shall be moving from the views very friendly to curiosity to those quite inimical to it.

Our first topics are the skeptics. In the long tradition of skepticism we shall be interested in two authors, Pyrrho (365/60–275/70 BC) and Sextus (ca. 160–210 CE) who, in his *Outlines of Pyrrhonism* (or *Skepticism*) claims he is presenting to the reader what are basically Pyrrho's views.[6]

His is perhaps the most puzzling discussion of investigativeness-curiosity, perhaps in the whole history of philosophy. He begins it by

[6] See Julia Annas and Jonathan Barnes (2000), *Sextus Empiricus: Outlines of Scepticism* (Cambridge Texts in the History of Philosophy), Cambridge University Press, second edition.

noting that Pyrrhonists carry on investigating and searching for the truth (I 3). The very name "skeptic" comes from *skeptesthai*, to investigate. Most importantly for us, *skeptikos* means "one who is disposed to investigate", "investigator", as Annas notes in her preface to the (Annas and Barnes 2000) edition of Sextus (p. xx). It comes very close to our "investigativeness", what we described as "active curiosity". Let us note that his is a case-by-case skepticism, where cases are "inked with real disagreement" (Williams 2010: 301).

Here inquisitiveness, our epistemic virtue (see the next chapter) plays the central role. The proposal is, however, not without its problems. The goal of investigation is, Sextus claims, *ataraxia*, the final well-being. If the investigator attains her final goal for which she is doing her work, she will stop. But then problems begin. A (rational) skeptic never stops investigating, so it seems that *ataraxia* is unattainable, the skeptic will never achieve well-being. On the other hand, *ataraxia* comes from investigation and it is not the case that investigation is the goal, finding is not.

But here the investigator faces the endlessness of possible cases to be considered: is this endlessness compatible with having reasonable goals, and attaining epistemic well-being?

Commentators have different reaction to this worry. Myles Burnyeat in his 1980 paper worries whether the skeptic can live his skepticism.

Christiana M. M. Olfert, in contrast, tries "to show that the Skeptic—or anyone engaged in a Skeptical investigation—arguably achieves a number of epistemic advancements or benefits when she achieves suspension of judgment. These, we might say, are the perks of Skeptical investigation (2015: 148).

She lists four epistemic improvements or advancements that may result from Skeptical investigation, all of which are connected with the Skeptics' stated aim of suspension of judgment (2015: 165):

First, "skeptical investigation benefits us by preventing the kind of cognitive failure that happens when we take false things to be true (or true things to be false)" (ibid.) (One wonders about it avoiding failure by not knowing anything is not much of a cognitive achievement.)

Second, "the Skeptic is in a much more sophisticated and informed epistemic position with respect to her puzzle after investigation than she was at the beginning, insofar as she now has a clear grasp of the reasons and arguments that are, and might be, given for and against various parts of the puzzle" (2015: 165).

A third epistemic benefit, she claims, "arises from the sensitivity and the precariousness of the state of suspension" (2015: 166) that blocks dogmatism.

Fourthly and finally, "skeptical investigation in that it promotes a life devoted to more, ongoing epistemic activity, which arguably has many of the same benefits as Plato's and Aristotle's notions of contemplation" (2015: 167).

I am giving this longer quotation to indicate the level of discussion which is going on these days on the virtues of investigativeness in the Pyrrhonic tradition. I am myself skeptical that the pursuit of happiness through investigation is compatible with complete lack of optimism about its results. However, I would like to point out that Pyrrho's view nicely illustrates one extreme concerning the amount of investigation we should perform in our efforts: the investigation should have no end! Of course, it refers to the project as a whole; in the matters of some particular issue (is this a computer in front of me?) I might come to the particular skeptical conclusion that I don't know the answer, and can't get to know it; not a very promising conclusion for a would-be whole-life investigator!

So much about radical skeptics. We now pass to Cicero, who is officially a moderate skeptic, which does not prevent him from praising knowledge and the search for it.[7] We can start with *De Officiis* [On

[7] For a fine reconstruction of his views see Raphael Woolf (2012), *Cicero: The Philosophy of a Roman Sceptic,* Routledge, in particular Chapter 2. See also Harald Thorsrud (2012), "Radical and Mitigated Skepticism in Cicero's *Academica*", in Walter Nicgorski ed. *Cicero's Practical Philosophy*, University of Notre Dame Press, pp. 133–151. Thorstrud points out that Cicero, on the one hand, wishes firmly to believe in certain 'dogmata' and, on the other hand, is a skeptic, aware that nothing can be proved:

These aspects of Cicero's character are reflected in his view of Academic philosophy. Cicero is forthright about the human epistemic predicament: just as it is supremely honorable to discern the truth, it is shameful to approve what is false as if it were true. His competing desires to avoid error and believe the truth feature prominently in his earliest as well as his latest expressions of allegiance to the Academy. As the only means of acquiring ever-closer approximations to the truth, Cicero must have thought the Academic method the best route to wisdom as well. He could hardly have offered it up to his fellow Romans as preferable to every other school if he thought otherwise. This is especially the case given Cicero's view of the great benefit he was providing his fellow Romans by

duties], Bk 1, Sect. 13, where Cicero lists specific qualities of human beings that distinguish them from beasts. One of them is that it "sees the causes of things, understands the rise and progress of events, compares similar objects, and connects and associates the future with the present—easily takes into view the whole course of life, and provides things necessary for it". The next one is: "/T/he research and investigation of truth/".

Cicero wisely adds: "/T/o this desire for seeing the truth (*veri videndi cupiditati*) is annexed a certain craving for precedence, insomuch that the man well endowed by nature is willing to render obedience to no one, unless to a preceptor, or a teacher, or one who holds a just and legitimate sway for the general good." We shall see in later chapters, in particular Chap. 9 on curiosity in science, that this combination is typical for many great scientists, from Newton through Darwin to Hawking: Cicero was indeed quite insightful in his description. His wider hunch is less persuasive: "Hence are derived greatness of mind and contempt for the vicissitudes of human fortune."

Here is another interesting point, presented as the advice he gives to us, his readers:

> In this quest of knowledge, both natural and right, there are two faults to be shunned,—one, the taking of unknown things for known, and giving our assent to them too hastily, which fault he who wishes to escape (and all ought so to wish) will give time and diligence to reflect on the subjects proposed for his consideration. The other fault is that some bestow too great zeal and too much labor on things obscure and difficult, and at the same time useless. (Sect. 18)

The contrast between two extremes characterizing desire for knowledge, hastiness versus too much labor, appears, in various guises, throughout the literature on curiosity.

In his *De Finibus,* Cicero gives fine examples of positive curiosity, fictional and historical. He interprets the offers given by Sirens in the *Odyssey* as offers of knowledge. And he contrasts passion for

---

encouraging them to engage in Academic philosophy. So for Cicero, the road to wisdom is paved with cautious opinion. (201: 134)

"miscellaneous omniscience" to the "contemplation of high matters": the latter is a matter of "a passionate love of knowledge" (XVIII, 48–9); again a contrast to be encountered throughout the history of writing about curiosity. The positive historical figures are Archimedes and Pythagoras (XVIII, 50), but also Plato and Democritus. He mentions love of fiction (XVIII, 52) as a positive instance of curiosity. So much about Cicero.

We now pass to Seneca, taking his well-known LXXXVIII. Letter to Lucilius titled "On Liberal and Vocational Studies' as our main source, since it gives the clearest picture of Seneca's views about the desire for knowledge (the elements of which Seneca retains throughout his opus).[8]

The domain discussed is already on the more appreciated side of possible objects of knowledge; it is the domain of liberal arts, highly praised by intellectuals. Seneca will proceed to narrow down the domain which deserves praise. Here is a contemporary reading of his goal, due to Martha Nussbaum:

> In his view an education is truly liberal [...] only if it is one that liberates student's mind encouraging him or her to take charge of her own thinking, leading to a Socratic, examined life and becoming a reflective critic of traditional practices. ("Cultivating Humanity", Liberal Education, series 1998, p. 40)

Seneca is adamant about his main claim: that only study that contributes to virtue has any value.[9] He talks about scholar investigating into language, lists activities like "pronouncing syllables, investigating words", and the like as paradigmatic and asks: "do such men teach virtue, or not?" If not, their knowledge, and their desire for it, are of little worth. Transposing the evaluation to present-day examples, we could guess, for instance, that Seneca would not highly esteem Chomsky's work in linguistics but would appreciate his virtuous political engagement. (Would

[8] As shown by Paulo Sérgio Margarido Ferreira in his paper on "Séneca e as artes liberais" (available at https://www.researchgate.net/publication/326196032_Seneca_and_the_liberal_arts).

[9] For the wider epistemological context of Seneca's reflections see Jula Wildberger (2006). For an interesting connection with the idea of self-knowledge elsewhere in Letters, see Margaret R. Graver (2014).

he then prefer Zizek to Kripke?) The same for the study of literature and music:

> Why try to discover whether Penelope was a pattern of purity, or whether she had the laugh on her contemporaries? Or whether she suspected that the man in her presence was Ulysses, before she knew it was he? Teach me rather what purity is, and how great a good we have in it, and whether it is situated in the body or in the soul.

So much about Seneca.

When we turn to Plutarch, the picture is completely different. First, we encounter a linguistic difficulty. His essay commonly translated as "On Curiosity" is titled "Peri polypragmosynes", and the word *polypragmosyne* has strong connotations of intrusiveness and of meddling into affairs of others.[10] Still, respectable interpreters, see, for instance, Van Hof, normally translate the title as "On Curiosity". Latin "curiositas" is sometimes used in the fully negative sense; for instance, Plautus writes that *curiosi sunt hic complures mali* (Stichus 198), "masses of evil people are curious"; see also Leigh (2013: 62). We shall be dealing with this problem in a moment, after we quote some Plutarch's typical opinions about the topic.

Plutarch is extremely critical of his object. The proper object of "curiosity" he talks about are evils, vices and misfortunes of other people, and the interest here is the desire to learn (*filomatheia*) about evils of others (*allotrion kakon*) (p. 335). The deaths of men, the shuffling off of life, "seductions of women, assaults of slaves, slanders of friends, compounding of poisons, envies, jealousies, shipwrecks of households, overthrow of empires are his typical examples" (p. 485).

What is the cure for such an evil? "Be less curious about people and more curious about ideas", proposed Marie Curie, indeed completely in line with what Plutarch would have said.[11] According to Plutarch, it

[10] For the wider epistemological context of Seneca's reflections see Jula Wildberger (2006). For an interesting connection with the idea of self-knowledge elsewhere in Letters, see Margaret R. Graver (2014). See also the excellent study by Matthew Leigh (2013).

[11] Maria Curie's anecdotal response to a reporter's inquiry. As quoted in Clifton Fadiman and André Bernard, *Bartlett's Book of Anecdotes* (2000), 150.

“shifting and diverting our inquisitiveness”, which loses its bad character once it changes its target. “Direct your curiosity to heavenly things and things on earth, in the air, in the sea”, he advises his reader:

> Are you by nature fond of small or of great spectacles? If of great ones, apply your curiosity to the sun: where does it set and whence does it rise? Inquire into the changes in the moon, and so on. (“On Curiosity” § v. (2007) *Plutarch's Morals*, p. 242)

Another relevant target is the reader him or herself. One should shift one's curiosity to oneself, and study one's own weaknesses, quite in Socratic tradition (515D–F).

One wonders about the nature and status of the shifted interest. It is, hopefully, not “meddling”, not *polypragmosyne*: so, what is it? Is it, after all, the good curiosity? Plutarch avoids a clear answer. One might think that he is indeed happy with the idea of good curiosity, but then the question remains of why he is never analyzing it, and why is he concentrating on the bad kind only. I propose the following understanding: for him, the paradigmatic kind of curiosity is meddlesome, and the central term to use for it is *polypragmosyne*, whereas the non-paradigmatic kind is what we see as normal curiosity. The other kind is worth mentioning only as a device for curing the bad, central kind (although he says at one occasion that such curiosity is laudable (517C–E, 518D, noted in Leigh 2013: 70).[12]

---

[12] Leigh goes much further in the positive direction, and associates the hypothetical positive stance of Plutarch with great positive changes in times to follow.

> The great change in usage from Polybius onwards is the association of polypragmosyne and its synonyms with the world of knowledge. […] from the Hellenistic period onwards there emerges a recurrent association of polypragmosyne with more innocent, often entirely commendable forms of investigation. This is apparent in all that has been written here of history, geography, and natural science and it underpins Plutarch's advice that we redirect the passion of curiosity from study of our neighbours' failings to the workings of nature or at least to the flaws and vices of an earlier age. Latin usage of curiosus from Cicero onwards reflects this understanding. (196)

I shall remain neutral on this further, very interesting possibility.

Augustine

Plutarch's condemnation of curiosity might have prepared the reader for a much more radical version, to be encountered in Augustine and present in his patristic followers. In his *On the Profit of Believing* (Sect. 22)[13] he distinguishes positive interest in things, "studiousness", from the negative one, curiosity:

> Thus as there is very great difference between one who studies any matter, and the absolutely studious; and again between him who has a care and the curious; so is there between him who believes and the credulous.

He then points to possible refinements: a man interested in his family, living abroad and inquiring about them would not be called studious; the term is normally reserved for interest in liberal cultural matters, not in one's own family. For our purposes, however, the crucial distinction is between the negative interest, curiosity and the positive one studiousness; exemplified each by a typical inquirer:

> [A]lthough both be led by great desire to *know*, yet the curious man seeks after things that no way pertain to him, but the studious man, on the contrary, seeks after what pertain to him.

These considerations from *On the Profit of Believing* seem to continue the Hellenistic and Roman line we already considered. However, there are passages in Augustin that are more radical. So, in his *Confessions*, he talks about "a certain vain and curious longing in the soul, rooted in the same bodily senses, which is cloaked under the name of knowledge and learning" (Chapter XXXV, 54)[14]:

---

[13] Available at https://www.logoslibrary.org/augustine/profit/index.html.

[14] We shall use *Confessions and Enchiridion*, translated and edited by Albert C. Outler (2006), The Library of Christian Classics, Westminster John Knox Press (July 1), https://www.ccel.org/ccel/augustine/confessions.xiii.html#fnf_xiii-p177.1.

> It comes from our appetite for learning (*in appetitu noscendi est*) and is therefore called "the lust of the eyes". (Augustine refers to Isaiah 14, but perhaps also to St John)

The main passage for our purposes is the following famous one:

> This malady of curiosity (*ex hoc morbo cupiditatis*) is the reason for all those strange sights exhibited in the theater. It is also the reason why we proceed to search out the secret powers of nature—those which have nothing to do with our destiny—which do not profit us to know about, and concerning which men desire to know only for the sake of knowing. And it is with this same motive of perverted curiosity for knowledge that we consult the magical arts. Even in religion itself, this prompting drives us to make trial of God when signs and wonders are eagerly asked of him—not desired for any saving end, but only to make trial of him. (Chapter XXXV, 55)

As many authors studying Augustin note, the passage implies that there can be bad curiosity concerning God himself. It seems that for him the right attitude in the case of religion is above all belief; any kind of questioning points to bad curiosity. Manson offers a useful summary:

> If we assume that God—and God alone—is the proper source of valuable knowledge, then we may manifest disrespect if we seek valuable knowledge in others ways. Or, if God has prohibited seeking knowledge of certain kinds, then to seek that kind of knowledge is disobedient. Worse still, if we assume that God requires us to adopt a range of commitments without evidence—then to seek the knowledge that would be evidence in support of those commitments is to under-mine trust in God. (2012: 243)

He also points out that this attitude survives in Christian classics all the way to Thomas Aquinas. He also tries to connect Augustine's line to non-religious contexts; we shall not follow him in these further efforts, since our task is not to make Augustine acceptable, but to locate him historically, in the great tradition that he initiated in Christian thought. But let me note that here again we encounter the importance of the distinction between the right and the wrong kind of epistemic caution or doubt, which we noted in Cicero; it will resurface again in relation to religious belief with James Clifford and William James, and is staying with us in our time.

To stay with Augustine for a moment, if one accepts the theological reading offered by Manson, as I think one should, one faces the question of the choice of belief. Augustine himself has gone from firm Manicheism to Christianity.[15] But how is one supposed to do this? You start as a Manichean; isn't any questioning of the doctrine on you part sinful, since it undermines a religious belief? Augustine might argue in an externalist fashion that the non-curiosity rule is valid only for the de facto true religion: the internal parallelism between questioning one's Manicheistic and one's Christian faith is simply not relevant. But this is not very persuasive, and also seems to fail in the situation of confessional pluralism that began to develop in Augustine's own time and flourished in the Middle Ages with a variety of Orthodox or semi-Orthodox churches and sects in the Middle East.

Unfortunately, we have to stop here, and leave the further history of curiosity in medieval times aside. Let me propose the take-home message for the whole section. The great thinkers of the Hellenistic period we talked about mostly turn to more human concerns, in contrast to high metaphysical interests of Plato and Aristotle (that did survive in Hellenistic time in the work of Neo-Platonists and started reappearing in early Christian literature). Their interest in curiosity is focused on strong evaluations, very often moral ones, of various kinds they individuate. Cicero is mostly positive about curiosity, but also initiates a question which will last for millennia: when you should stop being curious and investigating a give topic, and when you should go in depth, investigate and check. Seneca is less positive, and Plutarch seems to be obsessed with negative curiosity; again, the focus is on distinguishing various kinds of curiosity and offering and developing quite strong moral judgments concerning each.

[15] Manson himself mentions that in the *Confessions*, Augustine relates how he was interested in gossip and trivia, and in finding out about false Gods.

The Early Modern Positive Evaluation of Curiosity: A Minimalist Sketch

With modern times we witness the high positive evaluation of curiosity, like the one in Plato and Aristotle, but with a focus on a different, more scientifically minded curiosity. The birth of this modern positive evaluation of curiosity has been a long and complicated process. The literature stresses various aspects of it; in our time it avoids postulating a unique cause. Let me just sketch the five lines of development, with a lot of apologies for brevity. We have to be brief here, but the reader can find more in the chapter on scientific curiosity.

One line starts from geographical discoveries, above all the discovery of America(s): they showed, first, that there are important things to be discovered, not present in the available sources, and second, that the discoveries in various areas could be surprising, useful and epistemically impressive. The desire to discover, a sub-species of curiosity, can thus be legitimate. David Wootton in his history of scientific revolution (2015) points to "the "Columbus model" of the re-evaluation of discovery.

Another line supported the discoverer's curiosity and, stressed by Toby E. Huff in his (2011) book, stresses the invention of mechanical devices, "discovery machines"—like, for instance, and famously, Galileo's telescope—that made the research on nature easier and more attractive.

Let me illustrate. Huff lists "three contexts" that are for him "needed to understand the significance of the telescope" (2011: 18). First, he claims that it is "the instrument of empirical observation most associated with the scientific revolution of the seventeenth century" (ibid.). Second, "it transformed astronomy from a plodding science into an active, exploratory inquiry that constantly looks for new discoveries"; it became a "discovery machine". Finally, it dramatically enhanced the precision of observation.

Thirdly, thinkers-researchers start defending theoretical curiosity—from Copernicus and Bacon through Galileo and Descartes to Boyle and to the French Enlightenment and they gave legitimacy to such curiosity. Here are two quotes from Blumenberg's classical work:

> The early-modern renewal of the pretension to unrestricted theoretical curiosity turned against the exclusion of pure theory, and of the pure happiness that was bound up with it, from the realm of what could be reached in this world. (232)
>
> Copernicus became the protagonist of the new idea of science not so much because he replaced one world model with another, and thus showed by example what radical incisions into the substance of the tradition were possible, as because he established a new and absolutely universal claim to truth. Within the world there was no longer to be any boundary to attainable knowledge, and thus to the will to knowledge.
>
> The meaning of the Copernican claim to truth was admittedly only to appear and to be confirmed when Galileo and Newton, bringing mechanics to the aid of the anticipatory innovations in cosmology, sent Aristotelian physics into retirement. (1999: 361, Part III, "The 'Trial' of Theoretical Curiosity", in Hans Blumenberg (1966), *The Legitimacy of the Modern Age* [quotations are from the 1999 edition]).

Fourthly, a bit later, as a result a movement to re-evaluate curiosity positively, starts in academia, with disputations-dissertations defending curiosity. Even Church joins in.

Neil Kenny notes that curiosity was taken as a passion that is morally indifferent, but that in any given context manifests itself either positively or negatively (2004: 96). The virtuous or vicious character of manifestation depends on whether the subject and the object of curiosity are related in an appropriate way:

> Most often curiosity was defined as a passion and/or a desire, rooted in the body, morally indifferent in itself but always manifesting itself as a virtue or a vice. (2004: 96)

Fifthly, the fashion of collecting "curious objects" was born in early modernity, but it started weakening at the beginning of the eighteenth century and is arguably the least important of the five tendencies we listed.

Curiosity, in its science-oriented guise, has thus again become the central motivating virtue of human cognitive endeavor. Let me finally note that the development of physics in the scientific revolution culminated in the positive re-evaluation of understanding: the ambition to grasp the widest necessary connections between disparate phenomena, spectacularly implemented in Newton's synthesis, made understanding the central target of scientific curiosity. We shall say more about it in Chap. 9 on scientific curiosity.

So much about the great revolution of modernity. Our brief sketch cannot do any justice to the incredible richness of early modern apology of curiosity.

So, what has the historical work on curiosity done? We said too little to justify extended conclusions, but two results seem quite certain. First, it has stressed the importance of variations and richness of kinds of curiosity and has pointed to the fact that different kinds should be evaluated differently. Second, it has raised the question of the main epistemic role or roles of curiosity. To these questions we now turn in the topical part of the book, but without forgetting that a lot of topical reflection in fact stems from a dialogue with the rich history.

The Hellenistic interest in variations and kinds will stay with us for the next several chapters. We shall first discuss the main contrast, marking the work of thinkers like Seneca and Plutarch, the one between very bad curiosity on one side, and excellent and desirable one on the other. We shall defend the view that bad curiosity is still curiosity, only that it is not virtuous, following the already classical strategy of Philippa Foot for virtues in general. Then, we shall turn to the wide area of good (or not-bad) curiosity, which was central for Pre-Socratics, Socrates, Plato and Aristotle and is central in the contemporary discussion.

And we shall, of course, retain the interest in the main epistemic role of curiosity, and go as far as possible in characterizing it.

References

Annas, Julia, and Jonathan Barnes. 2000. *Sextus Empiricus: Outlines of Scepticism*, Cambridge Texts in the History of Philosophy. 2nd ed. Cambridge University Press.

Barnes, Jonathan. 1975. *Aristotle's Posterior Analytics*. Oxford University Press.

Blumenberg, Hans. 1966. *The Legitimacy of the Modern Age*. MIT Press.

Burnyeat, Miles F. 1981. Aristotle on Understanding Knowledge. In *Aristotle on Science: 'The Posterior Analytics' (Proceedings of the Eighth Symposium Aristotelicum)*, E.B. Padua, ed., 97–139. Reprinted in 2012 *Explorations in Ancient and Modern Philosophy Volume* II, 115–143. Cambridge University Press.

———. 2011. Episteme. In *Episteme, etc.: Essays in Honour of Jonathan Barnes*, ed. B. Morison and K. Ierodiakonou. Oxford University Press.

Fadiman, Clifton, and André Bernard. 2000. *Bartlett's Book of Anecdotes*. Boston, MA: Little, Brown and Company; Revised, Subsequent Edition (September 1, 2000).

Graver, Margaret R. 2014. Honeybee Reading and Self-Scripting: *Epistulae Morales* 84. In *Seneca Philosophus*, ed. Jula Wildberger and Marcia L. Colish, 269–294. Walter de Gruyter.

Huff, Toby E. 2011. *Intellectual Curiosity and the Scientific Revolution: A Global Perspective*. Cambridge University Press.

Kenny, Neil. 2004. *The Uses of Curiosity in Early Modern France and Germany*. Oxford University Press.

Kraut, Richard. 2006. The Examined Life. In *Blackwell Companion to Socrates*, ed. Sara Ahbel-Rappe and Rachana Kamtekar. Oxford: Blackwell.

Leigh, Matthew. 2013. *From Polypragmon to Curiosus: Ancient Concepts of Curious and Meddlesome Behaviour*. OUP.

Manson, Neil C. 2012. Epistemic Restraint and the Vice of Curiosity. *Philosophy* 87 (340): 239–259.

Nussbaum, Martha Craven. 1998. *Cultivating Humanity*. Liberal Education, Series.

Olfert, Christiana M.M. 2015. Skeptical Investigation and Its Perks: Diog. Laert. 9.69–70 and 79–89. In *Pyrrhonian Skepticism in Diogenes Laertius*, ed. Vogt Katja Maria. Tübingen: Mohr Siebeck.

Thorsrud, Harald. 2012. Radical and Mitigated Skepticism in Cicero's *Academica*. In *Cicero's Practical Philosophy*, ed. Walter Nicgorski, 133–151. University of Notre Dame Press.

Wildberger, Jula. 2006. Seneca and the Stoic Theory of Cognition: Some Preliminary Remarks. In *Seeing Seneca Whole Perspectives on Philosophy, Poetry and Politics*, ed. Katharina Volk and Gareth D. Williams, 75–102. Brill.

Williams, Michael. 2010. Descartes' Transformation of the Sceptical Tradition. In *The Cambridge Companion to Ancient Scepticism*, ed. Richard Arnot Home Bett. Cambridge University Press.

Woolf, Raphael. 2012. *Cicero: The Philosophy of a Roman sceptic*. Routledge.

Wootton, David. 2015. *The Invention of Science: A New History of the Scientific Revolution*. New York: Harper.

Yigit, Safiye. 2019. The Curious Case of Curiosity: A Virtue or a Vice? In *Toward New Philosophical Explorations of the Epistemic Desire to Know: Just Curious about Curiosity*, ed. Marianna Papastephanou, 150–165. Cambridge Scholars Publishing.

Zuss, Mark. 2012. *The Practice of Theoretical Curiosity*. Springer.

3

Nature and Kinds of Curiosity

Introduction

In this chapter we pass from the historical viewpoint to the topical one, keeping alive the themes we noted as being most prominent in the history of thinking about curiosity. The first task, favored in particular by Hellenistic authors, has been the inquiry into the nature of curiosity through inquiry into its various kinds. This topic will be central for the present chapter. Second, we have seen in the preceding chapter that one central issue addressed throughout the history of reflection of curiosity is the evaluative one, very closely connected to the first, taxonomical one: once you distinguish the kinds, you should evaluate them, often in terms of quite contrasting evaluations. We shall remain faithful to this project, doing it here in a brief form; the main evaluative discussion of curiosity as a virtue will be reserved for Part Two of the book.

Here is the plan. The first part of the chapter will be dedicated to the historically most popular divisions of curiosity. In the second part, we turn to one particular division that has been investigated in much more detail in the contemporary framework, thanks to Ilhan Inan, and then we conclude our picture of general and individual curiosity. The third part of

N. Miščević, *Curiosity as an Epistemic Virtue*, Palgrave Innovations in Philosophy,
https://doi.org/10.1007/978-3-030-57103-0_3

the chapter is dedicated to social divisions of curiosity; we would like to have a separate chapter on social curiosity, but it is impossible for reasons of space. This will then conclude our sketch of the nature and kinds of curiosity.

The Main Divisions

The General Taxonomy

The first distinction is relatively easy to draw, but once drawn, raises interesting problems. The motivation for inquiry may be intrinsic or extrinsic. Usually the term "curiosity" is used for the former, intrinsic case, although many authors speculate about further, extrinsic motives that drive what they call "curiosity". (Plutarch sees malice as the motive for meddlesome curiosity; Freud speculates about deeper and unconscious sexual motives; other examples will be mentioned a few pages below.) There is a substantive issue in waiting here. Consider an exchange between an employee conducting a task and his colleague:

> Why are you so curious about this stupid revolving mirror?
>
> Well, I am interested in producing a hologram to satisfy my boss who signed the contract, and I need to know about reflections of light from various angles produced in quick sequence.

Here, the final interest is purely extrinsic, and the interest in mirrors is extrinsic and instrumental. Still, one does count it in conversation as curiosity; it certainly is an epistemic interest functioning properly in scientific investigation. So, we shall treat intrinsic interest as a paradigmatic variant of curiosity, but accept the extrinsic instrumental one as a peripheral variant (peripheral to our interest in discussing epistemic virtue, not peripheral in general). Here is the division:

Goal	Extrinsic/intrinsic

This will help with normative reflection. The value of the practical curiosity obviously depends on the value of goals it is instrumental for. But having curiosity has been crucial to human survival; curiosity might have killed some cats and some people, but human life is unimaginable without it (or almost unimaginable; a trained philosopher can imagine things beyond any kind of possibility). So, given its crucial role in human survival, let us then assume it is at least in this respect mainly positive. So much for the moment about curiosity in general and the extrinsic subkind. As we said our main topic will be the intrinsically cognitive curiosity, of the kind often praised by philosophers.

Another interesting question which we can only touch here is the one we met briefly in Cicero. It is the contrast between two extremes characterizing desire for knowledge, hastiness versus too much labor, or contrast of optimism versus pessimism or caution; let me call it the issue of the style of interest. One can want to know, but prefer not to form the relevant belief, if there is a big chance of its being false; another person might rush a bit in order to achieve knowledge and risk ending with a falsity instead. The resulting extremes might be skepticism on the cautious end and unjustified dogmatism on the optimistic end. In short:

Style	Pessimistic/optimistic

The issue is often connected with the confrontation between William K. Clifford (publishing his paper in 1877) and William James (in 1886).[1] Michael Slote and Michael Williams objected in the discussion (my thanks go to both of them) that the cautious stance is not curiosity at all; however, I keep wondering why the cautious person would bother with caution, unless she were genuinely interested in arriving at knowledge. See also the discussion of when to halt one's curiosity in Fairweather and Montemayor, "Curiosity and Epistemic Achievement" in Watson et al. (2018), *The Moral Psychology of Curiosity.*

[1] William K. Clifford—"The Ethics of Belief", originally published in *Contemporary Review*, 1877; reprinted in William K. Clifford, *Lectures and Essays*, ed. Leslie Stephen and Frederick Pollock (London: Macmillan and Co. Clifford (1845–1879)—was an English mathematician. "The Will to Believe" is a lecture by William James, first published in 1896, 1886.

The next three contrasts are more studied by psychologists, and less by philosophers, so we shall return to them in the chapter on cognitive psychology of curiosity. One is the strength of curiosity, which can vary in intensity. The other is the mode: a person might be interested in the topic but passively waiting for some information to get produced, or she might be actively curious; a middle state between the two is also possible.

How long is a person going to be curious about some item? Psychologist are interested in factors that make an episode of curiosity into a short-term versus medium-term versus long-term one.

Finally, we often reflect about the epistemic (in contrast to moral) quality of our curiosity: how serious is my interest in contemporary dance, how much effort am I ready to dedicate to history of science and so on?

So, here is the table showing divisions concerning the quality of curiosity:

Quality	Strength	Mild/moderate/intense
	Mode	Passive/mixed/active
	Time	Short term/medium term/long term
	Epistemic quality	Low/medium/high
Quality	Level	Low/medium/high
	Goal	Extrinsic/intrinsic
	Style	Pessimistic/optimistic
	Strength	Mild/moderate/intense
	Mode	Passive/mixed/active
	Time	Short term/medium term/long term
	Epistemic quality	Low/medium/high

We now pass to another important criterion of division, the one concerning the target. This division has been central in contemporary account, in particular in Inan's book, and we shall dedicate more attention to it in a moment. One can be curious about some skill ("How does one drive bicycle?") or about more propositional and objectual matters. The first kind of target is knowledge-how; let us call the other "knowledge-wh-", to encompass knowledge-what, -whether and -why (plus some surrounding sub-kinds, like when). The next subdivision would contrast knowledge of some object, "knowledge-what" to all other, more propositional targets, the classical knowledge-that. I might be interested in my neighbor, in the town of Taipei, or in large cardinals, as opposed to being curious whether my neighbor is single, or why is Taiwan so successful in electronic industry. Similarly, an investigator dealing with the murder of JFK might rightly think: "The murderer of JFK is a serious criminal." Without having the least idea who the murderer is. Or, he can also have objectual knowledge, that it is Lee Harvey Oswald, and know enough about Oswald to individuate him.

There is a quite important further sub-division of curiosity targeting knowledge-that: on the one hand simple propositional target, on the other, connections-focused curiosity, aiming at understanding of connections and reasons and causes, expressed by appropriate why-questions (on appropriateness of questions see Whitcomb's 2018 paper). Curiosity has often been described as a desire for knowledge and understanding, and I think this may be the central kind of curiosity.

The stress on understanding might go back at least to Aristotle's texts we quoted, that is, *Metaphysics*, book Alpha and to *Posterior Analytic*, and to his crucial term "episteme". We noted that there is a respectable line of interpretation of it along the lines of understanding, defended prominently by Jonathan Barnes and Myles F. Burnyeat. We also mentioned the Pre-Socratic Leucippus, who allegedly said that "he would rather discover a single explanation (aitiologian), than acquire the kingdom of the Persians". (See the source in the chapter on history.)

Let me say a few more words about the contemporary understanding of understanding. Authors like Kvanvig (2003) offer a range of examples in which we would normally claim that the cognizer C understands some item.

Look first at objectual examples. For starters, one can say about C that he understands *his defeat* (why he lost); here, what is meant is that C is aware of the pattern of causes that lead to his defeat. Or, of C* that she understands *logic*, that is, grasps the patterns of dependence between logical items. If you say that C* understands *colors* you will probably mean that she grasps important relations between colors, and also the causal patterns in our color perception.

Next, an example of propositional understanding. John understands *that* he failed the exam suggest that John knows he failed, grasps the consequences, perhaps the cause(s).

And finally, understanding can be ascribed in the context of knowing-how, say, "Marina understand *how* to dance."

I shall assume that understanding is a species of knowledge, against the authors like Kvanvig (see his 2003 and 2012 works); for other authors, see the discussions in Stephen R. Grimm, Christoph Baumberger and Sabine Ammon eds. We shall be returning to the topic of understanding in chapters to follow, in particular in Chaps. 8, 9 and 11.

While we are at the target(s) of curiosity let me mention the contrast of scope: depth versus width. Again, one can go wide, in a disconnected, a bit chaotic manner, or in search of connections and unification; the latter option is more germane to understanding, and more valuable. As far as the mode is concerned we might distinguish the active variant (inquisitiveness) from the passive one featuring mere interest, too weak in the given situation to prompt active inquiry. We shall dedicate the whole next section to propositional, objectual and understanding-directed curiosity, appealing to the work of Ilhan Inan, who analyzed them in great detail.

The problems about intrinsic versus extrinsic curiosity arise again when we ask about the moral quality of the motivation: one often talks about sophisticated versus ordinary versus base (meddlesome) curiosity. The quality of motivation is often correlated with the level of the target: high versus medium versus low: from most ordinary and banal (Why are my neighbors quarreling? Why did Miss Ruritania cheat her boyfriend?) to most elevated (Why do basic cosmological constants have the values they have?):

Motivating object	Low ob./medium/high

This division has been central in the early history of philosophy of curiosity. Moralist philosophers like Plutarch give a rich menu of examples of low curiosity, similarly the religious thinkers like Saint Augustine. But note that one can have sophisticated interest in low-status matters; take a Sherlock Holmes–type detective, deploying his high intelligence to find out who stole something. The base, meddlesome curiosity is among the most often mentioned sub-kinds; but is it intrinsic or extrinsic? If it is extrinsic, it should not count against intrinsic curiosity, and certainly does not show that the latter is bad. I leave the question open, but note the conditional, as the basis for one possible line of defense of the virtuous nature of intrinsic curiosity against alleged counter examples. We have been dealing with this contrast at some length in the chapter on history of curiosity:

Target	Basic	Knowledge-how/knowledge-wh-
	Knowl.-Wh-	Objectual/propositional/ understanding
	Motivating object	Low ob./medium/high
	Scope	Depth/width
	Linkage	Disconnected/connected

Social Curiosity

We now pass to an interesting division, which, unfortunately, we shall not be addressing at any length in the book, except very briefly in the concluding chapter, for reasons of space; it would demand at least a chapter for itself. Until now, we have been formulating our divisions as if assuming that the bearer of curiosity is a curious individual. But there is an important alternative: the bearer can be a social entity, institutionalized or not. Think of schools and clubs, and of media thriving on curiosity of their consumers. This alternative would constitute an important topic for social epistemology:

Bearer	General	Individual/social
	Social	Non-institutionalized/ Institutionalized

The relevant areas are, for instance, law, media (cum politics), education and science (we shall say a few words on social epistemological issues concerning scientific curiosity in Chap. 9 on science).[2] The central matters concern inquisitorial cooperation, competition, division of labor and institutional transmission of information gained. Competition brings with itself a degree of vigilance that influence the style of socially organized curiosity. Division of work organizes the legitimate demands of curiosity.[3] Take the division of roles in schools between teacher and pupils, in sporting teams between a coach and the players, in law between agents of accusation, of defense and of judicial decision.[4]

Consider now some pretty obvious normative and value-related options. A particular sub-species of social curiosity or a given episode of its manifestation can be good or neutral or bad, in the purely epistemic, or moral, or prudential, or legal respect; finally, there might be some overall goodness or badness or neutrality belonging to it. Foucault's (originally Bentham's) panopticon is a good example of social curiosity that is bad in most of these respects.[5]

On the opposite side, the judicial organization in liberal, human rights respecting countries, testifies to methods that are good in epistemic,

---

[2] The pioneering work has been done here, as in neighboring areas, by Alvin Goldman.

[3] Linguists look at questioning in various social circumstances. An anthology by Alice F. Freed and Susan Ehrlich (2010), *"Why Do You Ask?" The Function of Questions in Institutional Discourse*, Oxford University Press, looks at asking questions in inquiry testimony, in police-suspect interrogations, in medicine, in a child-protection helpline, in the psychiatric assessment of patients, in teacher-student debates, in the workplace, in meetings, in genetic counseling and in broadcast journalism.

In each of these inquisitiveness plays a crucial role and the linguistic subtleties of questioning reflect the epistemological and interactional complexity of each type of inquiry.

[4] For the case of law see the comment by Goldman:

9.4 Common-Law Vs. Civil-Law Traditions

Modern adjudication systems are elaborate institutional structures. Labor is often divided among agents who uncover the evidence, agents who argue over the implications of the evidence, agents who issue verdicts on the evidence, and agents who oversee the entire process. Since the ultimate outcomes of a system hinge on the types of inter-agent transactions permitted and encouraged by the system, this is prime territory for social epistemology. The variety of possible adjudication systems is obviously legion, and some of these systems can be expected to outstrip others in their propensity to reach the truth. This signals a natural domain for veritistic epistemology.

[5] *Discipline & Punish: The Birth of the Prison* 1977 (New York: Pantheon). Originally published in French as *Surveiller et Punir*. 1975 by Editions Gallimard.

moral, prudential and legal respect. The presumption of innocence dictates cautious, Clifford-style curiosity on the side of all participants. Guilt has to be proved to very high standards, which brings epistemic benefits, prudential guarantees for citizens, and is morally crucial for the evaluation of judicial curiosity.

Similarly, in science the mutual control is crucial for the management of curiosity. Communication does counteract confirmation bias. Dan Sperber has offered a radical version of the thesis: purely individual cognition is biased and impartiality comes from communication. We don't need such a radical variant, but we are happy to accept that communication, if well organized, does counteract biases. (See Hugo Mercier and Dan Sperber (2017), *The Enigma of Reason*, Harvard University Press.)

Let us briefly mention the media. Here, the epistemic seductions are legion, and the promises of goodness either the purely epistemic, or moral or prudential, are scarce. Philosophers look for remedies:

> If we watch too much television that is probably a function of the fact that in our culture we do not spend much time training people how to use the medium: how to integrate it in a fulfilling life-plan. This, of course, may itself reflect an unwillingness in certain societies to include as part of basic education thinking about how one might lead one's life and the habits one needs to develop to pursue such lives. The problem of excessive television viewing, then, is a moral problem, if it is a moral problem, because of a larger cultural failing. It is not an ailment to be attributed to the medium as such, nor to its characteristic stylistic elaboration of the image. Neither the medium nor the image is inherently immoral, though our systematic failure to educate people about how to use it may be socially irresponsible. (Noel Carrol in M. Kieran (ed.), *Media Ethics*, Routledge, 1998, p. 151)

The manuals of investigative journalism extoll all the sources of goodness available. Take The investigative journalism manual published by Konrad Adenauer Stiftung (Singapore, 2016). Epistemically, they suggest digging deeply into an issue or topic, and providing understanding, not just superficial facts. Politically and morally, to be worthy of journalist's curiosity, the issue or topic has to be of public interest. And this will guarantee their prudential usefulness.

Let me then re-state the main normative points of view from which we can judge socially organized curiosity:

Practical normative and value status	Prudential	Low/medium/high
	Meaning-related	Meaningful/problematic/meaningless
	Moral	Bad/neutral/good
	Political	Bad/neutral/good
	Legal	Legitimate/problematic/illegitimate
	Overall status	Bad/neutral/good

Much more needs to be said, but we have to stop here. For a few more suggestions, see Conclusion, the section on social epistemology of curiosity.

Going Deeper into Kinds of Curiosity: Ilhan Inan and Objectual Curiosity

After this overview of the main kinds, we have to take a deeper step in the same direction. Here is why. As we said in Chap. 1, a few years ago Ilhan Inan produced the first philosophical book ever written specifically on curiosity, and in this book he raised some deep issues about two kinds of knowledge and the desire for knowledge that he characterizes as "propositional" and "objectual". His *Philosophy of Curiosity* is indeed a brilliant and original contribution to the discussion.[6] We have briefly mentioned the division of knowledge into objectual, propositional and understanding-constituted. Now we have to dig deep into this division. The contrast Inan is interested in is between these kinds of knowledge. Return to the example of the investigator is dealing with the murder of JFK. He rightly thinks: "The murderer of JFK is a serious criminal." Here, he operates with a proposition, and indeed he knows that the person is a serious

[6] Let me thank Inan on the long discussion we had. And let me note that one of the features that attracted me to Inan's account is his stressing the centrality of curiosity in people's coming to know what they do. On the first page of his book he tells us: "Hume called it 'that love of truth that was the source of all our inquiries'" (2012: 1). I will take the word "source" in a strong sense in the sequel.

criminal. But he does not know who the person is. His knowledge is indeed propositional, but it is not objectual, Inan would claim. Here is another illustration:

> Or suppose that someone is doubtful that there are any binary star systems and asks the question for this reason after realizing that the person he is talking to firmly believes that there are such things. Here the curiosity involved may simply be existential in nature (which is a type that I shall call "propositional curiosity") and could be expressed in terms of a definite description referring to a truth value or a fact, which I will have more to say in what follows. (2012: 49–50)

The bulk of his work in *Philosophy of Curiosity* is dedicated to the semantics and epistemology of objectual knowledge. Here is a typical short indication of a possible move concerning such knowledge (to be later developed almost to the size of the whole chapter):

> Propositional knowledge attributions are not fine grained enough to mark this difference. To know that p could happen in two different ways: one could know that p by merely knowing that the sentence expressing it refers to a fact, and another one may further know the fact it refers to. These correspond to what I will later call "ostensible knowledge" and "inostensible knowledge", respectively, in the next chapter. (54)

Here is one more illustration:

> If we hold that true declarative sentences refer to facts [...], then Holmes knows now that the sentence "Smith's murderer is insane" refers to a fact. He knows that the fact of Smith's murderer being insane exists. However, he does not know of any fact as being that fact. Just like it is possible for one to know that a definite description refers, without knowing its referent, Holmes knows now that a sentence refers without knowing that referent. To know the fact of Smith's murderer being insane more is needed. In order to know this, Holmes must come to know of a certain individual as being Smith's murderer and then come to know that he or she is insane. But he doesn't know this. He has merely inostensible knowledge of a proposition.

> As I discussed earlier, propositional knowledge attributions are not fi ne grained enough to mark this difference. (111)

> [I]t is a fact that a normal adult speaking a language has the concept of knowledge, not just in the propositional sense but also in the objectual sense, and is able to reflect on what he knows and does not know and acquire second-order thoughts. (134)

Throughout his book Inan scrutinizes some proposals from theories of reference with one or two great ideas from epistemological tradition, most importantly Russell's acquaintance/description contrast, in order, finally to arrive at his own original proposal. In short the initial and the final point should be characterized by considering the ways we refer to less known and better known objects. Here is the first and most important step in characterizing the contrast:

> Normally when we talk about something, there is a good sense in which we know what we are talking about. I may refer to the table in front of me, or to Socrates, or to Venus, or to the number 9 by using ostensible terms to pick them out. If I say "The table in front of me is wooden", I use the term "the table in front of me "to refer to a table that I have a clear vision of. I know what it is that I am talking about and referring to. In such ordinary cases the term I use to refer to the entity I wish to talk about is ostensible given that I know a certain object as being the referent of the term. Alternatively, I may use a term that refers to an unknown entity. Here are a few examples: "the population of Peru", "the oldest philosopher alive", "the nearest planet to earth that has water", "the last thing Russell said before he died", "the emotion that is felt the most by people". These are all inostensible terms for me now given that I do not know their referents. Now to get a feel for what I mean by an "inostensible term", a simple exercise is to ask yourself whether you know what these definite descriptions refer to. (2012: 33)

The main work done in the book is linking this contrast to traditional contrasts that dominate both the literature on mental reference and on acquaintance with external world. Here we have a hint about how to use the idea of definite description to characterize the partial and indirect grasp of the object, the grasp that characterizes the initial stage of the progress of curiosity:

> The paradigm case of an inostensible term is what Russell called a definite description (i.e., an expression that has the function of referring to a unique object). In English such terms usually start with the definite article "the". There are definite descriptions (such as "my mother") that do not contain the definite article but can easily be transformed into one that contains the definite article (such as "the mother of I"). Now what do I mean by a term being "inostensible"? In order to explicate this first we need to make a distinction between two ways in which a term may refer relative to a speaker. In the first case the speaker may know what the term refers to, in the sense that he knows a certain object as being the referent of the term, and in the second case one may lack such knowledge. Let us call the first kind of term relative to a speaker an "ostensible" term (for that speaker) and the latter an "inostensible" term (for that speaker). (2012: 33)

For Inan, curiosity is primary about objects. I shall call the Inan's stance "objectualism"; it suggests that in order to understand curiosity, one has to figure out how we refer to objects (topics) that we don't know sufficiently well. Having deployed a battery of fundamental distinctions from a theory of reference for descriptive-explanatory epistemological he arrives at the following definition of curiosity:

> Curiosity is a mental state that is the entertainment of an inostensible concept of something that is of interest to the subject. The entertainment of such a concept alone is of course not a desire, but it generally causes one. (2012: 133)

It should help us to deal with hard cases, for example, the case in which one is "in a state of total ignorance about the object of inquiry, and therefore "lacks any, even partial knowledge, or partial grasp, or partial acquaintance of it", but is still "in a position to start an inquiry and end it" (ibid.).

Let me discuss an important question starting from Inan's work. It has to do with his objectualism: for him, curiosity concerns knowledge of objects, in a wide sense of term. His analysis seems to me quite convincing, but it does seem to go against the grain of contemporary epistemology, where proposition and knowledge of them play crucial role. I would like to show that there is a route from objects to propositions, one in which understanding plays a pivotal role. The desire to understand is both objectual (Jane wants to understand the quarrel taking place in the neighborhood), and directed to the network of explanatory propositions (the neighbors quarrel because of this-and-this being the case). So, Inan's analysis can be usefully extended to the more usual propositional cases. The second question is taxonomical. The division of curiosity into objectual and propositional seems useful.

As we just noted, the object-focused knowledge naturally leads to knowledge involving propositional belief, and also that normative epistemology usually focuses on propositions (believed or known). But the objectual-propositional contrast remains. I try to show that the contrast can be made innocuous, and that the main bridge connecting interest in object(s) and the desire to believe true propositions is the desire to understand; the central case of curiosity. Even more importantly, it seems that knowledge demands some kind of ostensible acquaintance, a fact not sufficiently noted in the mainstream epistemology. Thus, Inan's story can be generalized to propositional cases and teach us about them.

Now, how does his objectualist account mesh with the usual normative insistence on propositional belief and knowledge?

A gap between the two would be a nuisance, especially if we believe that the descriptive and the normative matters normally fit each other, and if we see the point and interest of Inan's objectualist approach. Fortunately, there is a way to bridge the gap. The solution is the following: accept that acquaintance and other object-directed positive epistemic states play a major role, but note that acquaintance normally yields both propositional knowledge and a weakly holistic understanding. Similarly, the more ostensible knowledge is in many cases directly linked to understanding, and in others at least with some propositional material. So, where do we go (epistemically) from objects that are ostensible for us? Take Jane, a visitor to Turkey who is curious about the capital of Turkey; she knows there must be such a town, but the concept is not ostensible for her. Some of her questions might be focused upon a proposition, for example, "Is Istanbul the capital of Turkey? Well, no." She is still curious which town is the capital: Ankara is the capital. I agree with Inan that there are degrees of knowing what town Ankara is, but a lot of cases are such that simple propositional knowledge does satisfy one's curiosity and Jane's case might be one of them.

How is the objectual (aspect of) curiosity connected to knowledge, in particular to its central propositional variety? Curiosity is the desire to know; some philosophers claim knowledge is the primary and central epistemic item, others agree that it is at least of crucial importance, so the issue is a burning one. The first way to address it connects knowledge in general to the objectual curiosity, and the second connects the central variety of knowledge, namely understanding, to the objectual curiosity. (A reader might object that understanding is not a species of knowledge, following authors like Kvanvig (see his 2003 work); for such a reader the second strategy would not seem to work in a direct fashion. Still, some indirect route might be found, and I leave it for another occasion. Let me then pass to the first, most direct route.

Let me start with questions a curious person might raise. I shall rest content here with the simplest kind and with the easiest of options concerning them. Imagine Maria, very ignorant of geography. She is curious and asks: "What continent is Antalya on?" Suppose she thinks of Europe, Asia, America and Australia as exhausting the possibilities. A standard semantics of questions, nowadays called "alternative semantics", suggests that we can represents Maria's question by listing the alternative options, as an ordered quadruple:

<Europe, Asia, America, Australia>

and adding the imperative:

Tell me which one is the location of Antalya!

Sounds quite natural. But now assume that Maria is really, really ignorant. She has heard both of Asia and Eurasia, but she does not know which one, if any, is a continent, and doesn't know whether Asia is a continent or not (someone told her Asia is not physically separated from Europe, and she has no idea what the requirement for a piece of land is for it to be a continent, and soon. Then, the answer, "In Asia", is not sufficiently informative for her. Asia is not really a member of her sets of alternatives. This suggests deploying Inan's scheme: the concept ASIA is really not ostensible for Maria, and this is the problem. This suggests a criterion of adequacy: the specification of alternatives has to be in ostensible terms—so the ostensibility is a coordinate in the semantics of questions, coordinate orthogonal to the row of alternatives-plus-imperative. Now, this suggestion has consequences. Suppose a knowledgeable person who is an authority for Maria tells her the following: "Antalya is in Asia", and she believes that what he says is true. Does she know that Antalya is in Asia? Well, on the one hand, it seems that she can disquote the sentence from her belief-set: "'Ankara is in Asia' is true." and then come to belief and affirm that Ankara is in Asia. If she knows that "Antalya is in Asia" is true, she can come to know that Ankara is in Asia without much ado. On the other hand, if she is not clear about what kind of piece of land Asia is, if concept ASIA is not ostensible to her, one doubts that she really knows, in the full sense of "know" that Antalya is indeed in Asia.

On a commonsense and informal level, one would say that Maria's knowledge, if indeed she has any, is merely "verbal" and not "real". This points to a possible side-constraint on knowledge: in order to know that "A is F", one should have a sufficiently ostensive concept of "A". Sometimes, a minimal level is sufficient: if I am told by the right kind of authority that I trust there is a person called "Nuri Bilge Ceylan", then I know that Nuri exists. Name, plus kind-specification, namely that it is a real person are sufficient. If someone knowledgeable I trust tells me that Nuri Bilge Ceylan directed a beautiful movie (which is true), then I know it. But some minimal degree of ostensiveness is necessary (if I think that Nuri Bilge Ceylan is a park in Istanbul, I cannot come to know things about Nuri). If this is on the right track, then the non-ostensible/ostensible distinction can be, and probably should be, deployed both in the context of thinking about curiosity and in specifying requirements on knowledge. And, Inan's attention to the objectual turns out to be very fruitful; it is compatible with usual normative epistemology, but raises some very challenging questions for it. A lot more is to be said at this juncture, but we have to go on to the issue of understanding.

In understanding we encounter a combination of objectual and systematic propositional knowledge, often a network of propositions that gives one understanding. Take a very simple case mentioned at the very beginning of the paper: Jane wants to understand why her neighbors, a couple, are quarreling all the time. Suppose she learns that (1) the male neighbor is irrationally jealous of his partner, and (2) the partner, in turn, enjoys making him jealous but (3) hates the scenes he is making. Knowledge of the three propositions is a step forward. What is needed is to connect them in a causal network: she makes him jealous because she finds it enjoyable, but, she is irritated, because he is reacting so irrationally. And, all this leads to a vicious circle: this is why the neighbors are quarreling all the time.

I leave other cases of knowledge aside; let me just note that knowledge with analog content, as opposed to the propositional one, can yield a series of true and relevant propositions, thus resulting in important items of propositional knowledge. Let me illustrate. First the case of analog information. A student asks me: what did Descartes look like? I show her Descartes' portrait. Now she knows, but her information is not the classical propositional one. Nevertheless, she would normally be able to derive a lot of propositional information about Descartes' looks from the analog source. Second, I ask my sporting friend Boran right there, in the swimming pool: how does one swim butterfly style? Here, he says, and does a few elegant strokes. Fine, I start to receive a beginning of some knowledge-how; if I watch him swim with sufficient attention, I will be able to translate what I have learned in some elementary propositional information.

I conjecture that most interesting cases in which a single proposition is not enough to satisfy one's curiosity are cases where curiosity demands understanding. Given the centrality of understanding in our epistemic lives, there is no wonder that there is a long tradition linking it to curiosity. Inan mentions Descartes and Hobbes, whose respective definitions give pride of place to understanding. For Descartes, "[c]uriosity […] is nothing but a Desire to understand" (1989: 67). Inan also quotes Hobbes:

Hobbes does not care to philosophize on curiosity much, but at least he gives an explicit definition:

> Desire, to know why, and how, CURIOSITY; such as is in no living creature but Man; so that Man is distinguished, not only by his Reason; but also by this singular Passion from other Animals; in whom the appetite of food, and other pleasures of Sense, by prædominance, take away the care of knowing causes; which is a Lust of the mind, that by a perseverance of delight in the continuall and indefatigable generation of Knowledge, exceedeth the short vehemence of and carnall Pleasure. (T. Hobbes, *Leviathan*, ed. E. Curley, with selected variants from the Latin edition of 1668, 1994, p. 124) (2012: 1)

So, traditionally, curiosity is a desire for knowledge *and* understanding, or, if you accept my favorite picture, for knowledge and very often for its most relevant variety, that is, understanding, and it would be useful to keep the tradition alive in contemporary approaches.

Conclusion

Let us return to the general taxonomy of kinds of curiosity. Here is our whole proposal of taxonomy regarding (types of) curiosity, proposal of taxonomy of (kinds of) curiosity, both descriptive and normative, leaving the discussion of items not mentioned above for some other occasion:

General category	Narrow category	Dimension
Target	Basic	Knowledge-how/knowledge-wh-
	Knowl.-Wh-	Objectual/propositional/ understanding
	Motivating object	Low ob./medium/high
	Scope	Depth/width
	Linkage	Disconnected/connected

General category	Narrow category	Dimension
Quality	Level	Low/medium/high
	Goal	Extrinsic/intrinsic
	Style	Pessimistic/optimistic
	Strength	Mild/moderate/ intense
	Mode	Passive/mixed/active
	Time	Short term/medium term/long term
	Epistemic quality	Low/medium/high
Bearer	General	Individual/social
	Social	Non-institutionalized/ Institutionalized
Practical normative and value status	Prudential	Low/medium/high
	Meaning-related	Meaningful/problematic/ meaningless
	Moral	Bad/neutral/good
	Political	Bad/neutral/good
	Legal	Legitimate/problematic/ illegitimate
	Overall status	Bad/neutral/good

Let me just offer two toy examples, of how the characteristics we listed, combine in typical cases of curiosity, individual and collective. Take individual's ordinary social curiosity, as exemplified in watching TV news. The knowledge pursued concerns objects (I want to know about the Ukraine), then knowledge-wh-, and finally, and very importantly, understanding of the situations, moves of the agents in the situation, and so on. The motivation is often intrinsic, mode active but not too active, the style mostly optimistic, level low to medium, and the temporal dimension typically the short-term one. Ordinarily the scope tends to breadth, rather than to depth, and the normative status is most often neutral or good. Compare it to the bad low-level curiosity of gossipers, again hunting for propositional and objectual knowledge and for understanding. Often the goal is intrinsic, the mode (unfortunately) active, style optimistic (with little or no checking for error and disinformation), the scope going into width, and temporal dimension is the short-term one. Often, the normative status is bad.

Pass now to professions and consider the kind of curiosity expected by, say, a defense lawyer. First, in regard to its object it is curiosity concerning whether, why and how, encompassing propositional focus, and objectual focus, and, in the center of it all, understanding of the case at hand. The required motives are extrinsic, in the first place the welfare of the client, the expected mode is active (the more active the better). When it comes to the evidence, the required style is highly vigilant (epistemically "pessimistic"). The level is high, the scope (relative) depth, and the temporal dimension a long term one. Finally, on the value side the social expectation is that the curiosity is hopefully of the good, virtuous kind. Compare this set with the expectations from a typical surgeon team. The knowledge required is above all high-level know-how, the primary motive again extrinsic (the patient's health), mode, active, but style optimistic in contrast to the defense lawyers. Temporal dimensions vary: short term, concerning a particular patient; long term, more generally. Scope is again depth, and the value very positive.

Hopefully, the proposed dimensions offer a potentially useful grid, and their combination offers appropriate pigeon-holes for classifying typical motivation for commonsense types of intellectual (or cognitive) activities, individual or social-collective. Moreover, the optimist/pessimist contrast, combined with others, helps situate historically important types of skepticism and dogmatism (and perhaps types of confusion, like postmodernist rejection of truth as the goal of inquiry). On the side of social epistemology, the classification can (and should) be extended to motivation in transmission of knowledge, with applications to testimony, and teaching.

Here is the last hint. The study of possible improvement of our knowledge should address the issue of road from less valuable to more valuable within some of the dimensions, or even from one neutral extreme to the other. How do people pass from passive curiosity to active inquisitiveness? How do they slide from extrinsic interest to intrinsic ones? Psychologists talk about the functional autonomy of motives: a motive that started as instrumental-extrinsic can take over and become a goal in itself. Another question for educators is how to get from low-level interest to middle and high? Again, what is the road from knowledge-that to knowledge-how and -what for the other way around? Finally, an important issue for social epistemology is the road from individual to social or vice versa. So much about taxonomies of curiosity.

Let us conclude Part One. In the bulk of the text we have presented the main proposals made in the early history of the notion of curiosity and love of knowledge, both in their historical and in their topical aspect.

We now pass to evaluative matters. Is curiosity an epistemic virtue or a vice? These matters will form the central part of the book.

References

Inan, Ilhan. 2012. *The Philosophy of Curiosity*. Routledge.

Kvanvig, Jonathan L. 2003. *The Value of Knowledge and the Pursuit of Understanding*. Cambridge University Press.

———. 2012. Curiosity and a Response-Dependent Account of the Value of Understanding. In *Knowledge, Virtue, and Action: Putting Epistemic Virtues to Work*, ed. T. Henning and D. Schweikard, 151–164. London: Routledge.

Whitcomb, Dennis. 2018. Some Epistemic Roles for Curiosity. In *The Moral Psychology of Curiosity*, ed. Ilhan Inan, Lani Watson, Dennis Whitcomb, and Safiye Yiğit, 217–237. Rowman & Littlefield International Ltd.

Part II

The Central Role of Curiosity

4

Is Curiosity a Virtue?

Preview

We have seen that in the history of the debate one central question was whether curiosity is a virtue or not, and if yes, what kind of virtue. Virtue is again the fashionable topic of the day, both in ethics and in epistemology. In this book I am joining the fashion and treating curiosity as a virtue; in fact, I am calling my proposal *the* motivating virtue account,

This chapter will provide general explanations and arguments in favor of treating curiosity as a virtue. For the beginning, we need a picture of virtue, and of the general considerations that go with it. Then, we shall address the main historical argument against curiosity being a virtue, namely cases of really bad curiosity, like the ones we encountered in Plutarch. We shall appeal to some contemporary proposal of how to solve the problem of bad candidates without compromising the kind to which they belong.

In contemporary discussion, much more attention is payed to mild vices of curiosity, like superficiality; we shall discuss these in the next two chapters.

N. Miščević, *Curiosity as an Epistemic Virtue*, Palgrave Innovations in Philosophy,
https://doi.org/10.1007/978-3-030-57103-0_4

Being Virtuous: The General Considerations

What Is a Virtue?

Let us start with proposals concerning the nature of virtue. In his *Nicomachean Ethics*, Aristotle talks about "a purposive disposition, lying in a mean … and determined by that which a prudent man would use to determine it" (ii.6, 1107a1–2). He sees it as an excellency, and I shall assume with most writers, from him to Axtell, that virtues are "excellences of the agent".[1]

The main distinction proposed in the literature is then between "executive virtues" having to do with good functioning of our capacities, from character-linked ones. The latter normally motivate its bearers to act in accordance with them.

Now, in the literature, the distinction is combined with the ethical/epistemic contrast.

Ethical virtues like fairness, generosity and courage normally motivate its bearers to act in accordance with them. For many authors a necessary condition of a trait being a virtue is that it at least partially contribute to, or constitute, the flourishing (*eudaimonia*) of the possessor of the virtue.

Epistemic or, more narrowly, intellectual virtues are apparently of two kinds. Some are motivating character-traits, similar to ethical virtues; intellectual courage, integrity and intellectual humility are good examples. Other items, famously counted by Aristotle as intellectual virtues,

---

1

The first part of what is now the present chapter has been presented at the Sofia conference on 15 June 2007. I have profited a lot from the very rich discussion, for which I am thankful to the Sofia philosophical audience, especially Elisabeth Fricker and Marina Bakalova. Other parts have been discussed in blogs, and I am thankfully referring to my correspondents in the text. General thanks go to J. Kvanvig, G. Axtel and D. Pritchard for blog-hospitality and encouragement. Finally, a great debt is owed to Ernest Sosa; first, in terms of a long-term inspiration with his work and, second, for discussion and criticism. I see my proposal somewhat immodestly as extending his own and supplementing it with a more general framework.

For the discussion of Foot's solution thanks go to my curiosity class at Central European University (CEU) 2020, and in particular to Duje Kovacevic; the discussions we had helped me shape the final version of my own proposal.

for instance, intuitive reason, are just capacities. Sosa has proposed the full list of cognitive capacities as his candidate virtues. However, a worry arises. Thomas Aquinas has already noticed the fact that virtues from the Aristotelian list, unlike the moral ones, do not motivate, and his worries naturally generalize to Sosa-style list of all virtues-capacities. Such virtues are not "operative" he says, and they do not "perfect the appetitive part, not affect it in any way"; that is, have nothing to do with "appetite" for reaching cognitive goal(s). They might be called "virtues", but only "in so far as they confer aptness for a good work, viz. the consideration of truth" (see the footnote for sources).[2] This issue, call it "Aquinas' problem", brings us to two central choices, or dilemmas, of VE: either stress the motivating, moral-like and moral-related virtues, or rest content with those that merely confer aptness for the consideration of truth without motivating it.

The authors who choose the first horn often see themselves and are seen by others as being pretty much of revolutionaries within epistemology. L. Zagzebski is a leading figure in this camp, and she is interesting in bringing the full panoply of moral virtues to bear upon epistemological issues. Writing about her and her followers, Brady and Pritchard speak about "the radicals who maintain that this renewed focus on the virtues can actually offer us an entirely new contemporary epistemological theory/ethical theory".

---

[2] Aquinas diplomatically avoids the issue in his comment to the relevant chapter (Book E) of *Nicomachean Ethics*, but raises it in *Summa Theologica*:

> It would seem that the habits of the speculative intellect are not virtues. For virtue is an operative habit, as we have said above (55, 2). But speculative habits are not operative: for speculative matter is distinct from practical, i.e. operative matter. (I./II, q. 57)

and then answers:

> Since, then, the habits of the speculative intellect do not perfect the appetitive part, nor affect it in any way, but only the intellective part; they may indeed be called virtues in so far as they confer aptness for a good work, viz. the consideration of truth (since this is the good work of the intellect): yet they are not called virtues in the second way, as though they conferred the right use of a power or habit. (Ibid.)

The other camp is led by Sosa and Greco, who develop the reliabilist tradition, in the direction of including virtues in the account of knowledge. However, their virtues are non-motivating virtues, either capacities or features like cognitive integration that confer aptness for the consideration of truth and belong squarely to the Aristotelian tradition. Brady and Pritchard characterize them as "the conservatives who simply urge that we should place discussion of the epistemic/moral virtues at the center of epistemology/ethics, but who make no claim to be offering a wholly new contemporary epistemological theory/ethical theory. Instead, such theorists just regard themselves as refining views that are already common currency in the current literature".

Here is John Greco's summary of the distinction:

> What is a virtue in epistemology? In the broadest sense, a virtue is an excellence of some kind. In epistemology, the relevant kind of excellence will be "intellectual." But then what is an intellectual virtue? Some philosophers have understood intellectual virtues to be broad cognitive abilities or powers. On this view, intellectual virtues are innate faculties or acquired habits that enable a person to arrive at truth and avoid error in some relevant field. For example, Aristotle defined "intuitive reason" as the ability to grasp first principles, and he defined "science" as the ability to demonstrate further truths from these. Some contemporary authors add accurate perception, reliable memory, and various kinds of good reasoning to the list of intellectual virtues. These authors follow Aristotle in the notion that intellectual virtues are cognitive abilities or powers, but they loosen the requirements for what counts as such.
>
> Other authors have understood the intellectual virtues quite differently, however. On their view intellectual virtues are more like personality traits than cognitive abilities or powers. For example, intellectual courage is a trait of mind that allows one to persevere in one's ideas. Intellectual open-mindedness is a trait of mind that allows one to be receptive to the ideas of others. (Greco, John (2002), "Virtues in Epistemology", *Oxford Handbook of Epistemology*, pp. 287–315)

We shall appeal to both kinds of epistemic-intellectual virtues in discussing curiosity. To anticipate a bit, we shall take curiosity in a non-problematic fashion as a character trait and then argue that it is a positive character trait, an excellency. Then, in the next chapter, we shall argue that its main function is motivational, namely, to motivate and organize intellectual virtues-capacities.

We cannot appeal to authorities here. As we noted, "curiosity" has stayed at the margin of epistemological debate. Even the virtue turn did not make it very popular. Fortunately, there are exceptions, like Baehr (2011) and Brady, whose ideas will be mentioned and discussed in the chapter. Another valuable exception is Roberts and Wood's 2007 book *Intellectual Virtues*; its authors see curiosity, or "love of knowledge" as they call it, as a presupposition, or "necessary background of the other virtues". The love of knowledge "provides the intellectual motive for exemplifications of such virtues as humility and courage" (2007: 306). But if it motivates the exemplification of other virtues it should be foregrounded, not treated as "background". Curiosity seems to be something of a character trait; but character-based virtue epistemology did not concentrate much upon it.

Let me say a few more words about the two paradigms. For starters, let us note that there are important sources of tension within the virtue epistemology paradigm. Epistemology is traditionally concerned with knowledge and stresses the importance of truth, of one's reliable arriving at truth and being able to tell oneself whether one is thus reliable. Truth is central, whereas the positive qualities, "virtues", of the cognizer are just means to arrive at it. The view that takes this traditional and quite plausible view of the enterprise of cognition seriously, takes virtues as being important but not as being fundamental. As we just noted, Ernest Sosa is the pioneer of this approach (see Sosa 2007). Naturally, the virtues involved are more like executive qualities than like traditional moral virtues that primarily *motivate* the agent to pursue her goals, and only secondarily aid it in pursuing it. Further, the central role of success in reaching truth stands in stark contrast to the relatively modest role of success in traditional virtue ethics, which concentrates more on the right motivation and intention plus honest trying than on external success.[3]

On the other hand, a serious involvement with virtues, in particular moral ones, would prompt, and has prompted philosophers to look for motivating, moral or moral-like virtues in the domain of cognition. They might even propose that there are "ethical foundations of knowledge", as the subtitle of Zagzebski's (1996) book *Virtues of the Mind* suggests. But epistemology does not seem to be just a branch of ethics. So, it seems that truth and virtue don't go well together in epistemology, and that if you maximize the value of truth, you minimize the one of virtue, and the other way around. Moreover, even most of the other proposals within this camp, except Zagzebski's one, stop short of claiming that the values pursued in epistemic enterprise are themselves dependent on virtues. Rather, virtues enable their bearers to realize or implement the values independent from virtues themselves. Indeed, how could virtues ground or produce the value of truth, or understanding? So, even in this "moralizing" camp one rarely finds a virtue-*based* approach. Most philosophers in both camps are thus content with mere virtue-focused line and a virtue-based project is a bit too much for them.

This brings us to our main topic. *If* curiosity is a virtue, epistemically it is a very important one, endowed with a great motivational power. Consider other virtues. Some of them, central for authors like Sosa and Greco, are mainly abilities, like intelligence and inferential abilities. They don't by themselves motivate their owners to inquire (Sosa notes this, and in his recent work adds additional factors; we shall discuss it briefly in Sect. 3). Others, closer to their moral counterparts, say intellectual humility and courage, motivate a particular, virtuous behavior in the situation of inquiry (and in situations of passive reception of information). Take the example of a brave investigative journalist, who manifests intellectual courage in her work; we naturally assume, given difficulties and risks of searching in the area, that she was extremely inquisitive, and extremely motivated to find out. Or take epistemic humility: a student shows her

[3] As Julia Annas has pointed out in her 2003 book. She also mentions another main dis-analogy between intellectual and moral virtues, as against Zagrebski's idea of "ethical foundations of knowledge": even if we treat moral virtues as skills, as Socrates and Stoics do, and thus bring them closer to the intellectual ones, the aims of the two groups are too different for a unified account.

professor that he has made a stupid mistake arguing for a conclusion (this actually happened to me the day before I am writing these lines). If the professor is intellectually humble, he would gladly accept her correction. Well, the assumption in the humility case is that he would do this because he is interested, at least partly intrinsically, in whether his conclusion is true, and his moves correct; again, a motivating force distinct from humility itself is required, and the motivating force has to do with interest in how things are. So, character-related epistemic virtues normally demand some kind of curiosity, in order to start functioning at all. (In the epistemically non-central cases, those of almost completely passive reception of information, they demand for their exercise at least caring for truth and for avoidance of error). Now, *curiosity offers this motivational underpinning, providing the epistemic goal for virtues-abilities and the opportunities needed by character virtues.* Therefore, *if curiosity is a virtue, it is motivationally basic.* Further, within virtue framework, items that are motivationally basic are basic *tout-court,* in contrast to merely executive abilities, and to items, like character virtues, that do motivate but only within a framework offered by the basic motive. (I leave aside for the moment the practical, extrinsic motives; they are certainly not basic in any epistemic sense). So, *if curiosity is a virtue it is a basic one.* As I said, I call my account of curiosity based on this idea "the motivating virtue account".

Plutarch Steps In

Of course, there is a catch: the little "if" that makes it all conditional. As we saw many philosophers thought curiosity is not a virtue. We noted that Plutarch describes it in his "On Curiosity" in completely negative terms, as an "unhealthy and injurious" state of mind which allows "winter and darkness to enter the soul". He is concerned with low-level active curiosity, to stay with our division from above; mainly with curiosity about the whereabouts of other people. His argument for badness of such curiosity seems to be, in the nutshell the following: a curious person is interested in what is hidden, since things in sight need no special inquisitiveness to be seen; what people hide are their weaknesses and misfortunes, therefore the curious person is curious about these. But typical curious neighbor is not a Mother Theresa type, interested in misfortunes in order to alleviate them. Therefore, the only reason he (Plutarch is nicely politically correct about ascribing this bad curiosity to man) can have for his inquisitiveness, is maliciousness; and this then makes curiosity extremely bad. Well, perhaps he did not mean curiosity as we know it, but just meddlesomeness as his Greek suggests (the word he uses, "polypragmosyne" is normally used just for the bad variant of curiosity). But others were clearer: Pascal claims we want to get to know things only in order to brag around with our knowledge, and Montaigne, in his more pessimistic moments comes quite close to the same kind of judgment.

Let me note that the linguistic issue might be more general. The word "curiosity" is often used as synonymous with "nosiness" whereas, the grand sounding "thirst for knowledge" is usually deployed in a positive sense. We need here a very general term, so we shall use "curiosity" to cover all kinds of cognitive interest. We should take into account varieties of curiosity; not all of them have the same status, and if the central ones are normatively good (or bad, or neutral), then this will be a sign that curiosity is to be treated as a good (or neutral or bad) thing. Let me also note that the normative status of curiosity has several dimensions: an occurrence (or a sub-kind) of curiosity might be good or bad in the purely epistemic sense, in the prudential sense, in the moral sense, further as being meaningful or meaningless, and finally in its overall character. In the social context one may add the legal and political status of a certain sorts or occurrences of curiosity (such as scandals coming from the web sources and the kind of issues they are raising).

The Problem of Bad Curiosity

Introduction

Challenges to inquisitiveness-curiosity being a virtue are legion. In his *Rules*, commenting on Rule One, Descartes debates with Montaigne the claim of the latter that one should be curious only about "one's business", things that concern oneself, and are hopefully in one's power. Descartes replies that we can never know where our inquiry will lead us, and symmetrically, what things are going to be our business. Maybe, the use of "inquisitiveness" instead of or together with "curiosity" could help, and this is how I will refer to it here. The main objections focus on two points. First, that there might be cases of trivial, imprudent and immoral inquisitiveness-curiosity. This objection from triviality has been put to me in discussion by Sosa, who then used as example the curiosity about details of history of chewing gum, while in the published work he mentions curiosity about the number of blades of grass and other quite humorous examples (2001: 40). Second, and more importantly, objections from imprudent and immoral cases are quite natural and they make up the core of Plutarch's and Augustin's cases about ordinary curiosity. The reservations are still alive today, for example, they have been put by to me by *Jonny Blamey* in the blog discussion (*Certain Doubts,* August 14th, 2006):

Another problem is that curiosity doesn't seem to be a pure virtue. Just survey some folk wisdom "Curiosity killed the cat", "Don't be nosey", "Mind your own business". Privacy is valued, and therefore there is a certain amount of moral approval in respecting others' privacy. Although curiosity in a student is a sought after and valued quality, the sort of sordid curiosity that spawn the paparazzi and salacious gossip mongers is usually thought to be far from commendable. In fact there is a strong, though perhaps contemporarily unfashionable, idea that innocence is a virtue. Children shouldn't know certain facts, and therefore their curiosity in certain areas should be curbed. Libertine experimentation is driven by a curiosity of a most unsavory kind.

Let me mention a contrast within the seriously bad cases, noticed by Roberts and Wood (2007: 156). Compare two persons for whom we would say that their curiosity is far from positive. John is systematically interested in unimportant matters, and spends hours investigating trivial question he cannot answer. Fred is systematically interested in morally deeply problematic matters, say the intimate details of his neighbors' sexual lives. John's curiosity is cognitively deficient, Fred's is morally deficient. We shall take both kinds into account.

In general, virtues can sometimes produce terribly bad results, and curiosity is no exception. To illustrate, here is the kind of example that can be used to illustrate various points about virtue. Consider the following imagined situation from a concentration camp: every time a virtuous camp inmate shows compassion (generosity, etc.), the SS camp guard, unbeknownst to her, tortures and executes 100 inmates. Her compassion is not very helpful, and, due to the guard's malice, it "produces" terrible consequences each time it manifests itself. With a bit of bad luck, any moral virtue can get counterproductive. So, why wonder about the same thing happening to epistemic virtue?

Defending Curiosity

First, a general answer. The examples adduced, together with a host of other one can think of, seem to point to the conflict of values and conflicts between various virtues. This is usual for virtues on ordinary understanding: famously, justice might conflict with mercy. The same holds for inquisitiveness. It can conflict with moral and with prudential interests. Then it does not always stop being a virtue, but it can be overridden. Sometimes it is overridden, and then we use the saying that curiosity killed the cat. Sometimes it is not. "Curiosity killed the cat, but satisfaction brought it back" wrote Eugene O'Neill. (Unfortunately, it is a quote found on and by Google that I could not locate more precisely.) Some bad consequences of curiosity clearly do not make it non virtuous. For example, curiosity killed many brave researchers, like, for instance, Marie Curie, and we don't think that she was non-virtuous as researcher because of this; quite the contrary. We should distinguish and look at particular cases.

Here is then our plan. We see truly bad curiosity as a worrying counterexample to the view that curiosity is a virtue: we find the less problematic cases, like possible triviality of some instances of curiosity less worrisome. So, we shall concentrate here on the truly bad kinds, and look for a solution. Once we find one, in fact borrow it from the extant discussion of virtues in general, we shall proceed to our positive story. This will be the topic of our next chapter on curiosity as the central motivating virtue, in fact the central chapter of the whole book. After that, we shall return to less problematic weaknesses of curiosity, and discuss objections based on triviality, superficiality and irrelevance. We hope that once we solve these, the idea that curiosity is the central motivating epistemic virtue will show all its attractiveness.

Here we shall do the most difficult job, consider the cases of dramatically bad curiosity, of Plutarch type, and their strength in possible besmirching of curiosity and depriving it of the status of a virtue.

In this and the next chapter we shall go much further in our praise of curiosity and argue that it is the force that is normally mobilizing and organizing, or helping to organize epistemic virtues, both character-virtues and abilities. Of course, we are inquisitive, curious and alert partly because of practical interests, searching the means for practical ends, partly because of pure need to know, or, from both at the same time. Our inquisitiveness-curiosity is thus either pure or practical or mixed.

So, if I am curious about what time it is, since I have to come on time to the meeting, my curiosity is instrumental, and the final goal, coming on time, is the main factor in explaining my interest. Still, curiosity is doing some work: imagine my counterpart who wants to come to the meeting, but simply cannot bring himself to inquire what time it is: his way of thinking and deciding would be seriously deficient.

Here we shall concentrate upon the pure variety. A human being devoid of curiosity would have little motivation to arrive to true belief and knowledge. On the usual view of motivating virtues, this would seem to make curiosity into a virtue; since it is the mainspring of motivation, we should take it as the motivating epistemic virtue. This argument will be the main topic of the chapter.

. Let me remind you that in the foregoing chapters we announced our taking the notion of inquisitiveness-curiosity in the widest sense, so as to encompass primitive alertness to the features of the environment (again, if this doesn't fit your semantic intuitions, then take it, please, as a stipulation, not as analysis of the commonsense notion). Alertness is biologically based, as shown by the existence and functioning of early warning perceptual mechanisms, that alert the organism to potentially threatening changes in its surrounding, and by the mechanism of habituation that makes it "loose interest" in repeated stimuli. The importance of this wide sense of inquisitiveness is that it helps us address the problem of cases of "low-grade knowledge that are unaccompanied by any genuine motives or actions" as Jason Baehr characterizes them (2011: 44). The answer is to take alertness as proto-curiosity, and in this sense, a proto-virtue, that makes us open to the world and sensitive to it.

We need one more widening. As several colleagues have reminded me, a person, finite or infinite, who knows everything would not be curious, and would thus paradoxically lack the alleged main motivating epistemic virtue. One answer is that many human virtues, like courage and generosity, are tailor-made for human agents in less-than-perfect but better-than-hellish human circumstances. Curiosity is one such virtue, typical for finite and relatively ignorant beings, in need of constant updating of information in order to function successfully. But I prefer another line: I will just stipulate a slightly wider meaning of "inquisitiveness" that also includes cherishing the truth once found. It seems to me a natural extension of meaning: a person with bad memory but eager to get to know, who subsequently doesn't care a bit for the knowledge acquired and is completely unworried about having forgotten everything she learned, is not consistently inquisitive. So, the hypothetical omniscient person who keeps her virtue by cherishing what she knows is "curious" in this wider sense.

Duje Kovačevič has raised the question here: Is the omniscient person prone to having bad memory or is she epistemically perfect? Because if an omniscient person is epistemically perfect, at least in the sense of not being able to have bad memory (or at least not so perfect memory), then she is by default unworried about forgetting what she knows—because she cannot forget. In short, if A cannot forget what she knows, then she is by default unworried about forgetting (because forgetting is not possible for her). But would that mean that the person is not consistently inquisitive? Hopefully, this is not a problem for my proposal: I mention careless forgetting as an example of non-cherishing, not as an act of omniscient person.

To repeat, we have been arguing that curiosity, in spite of problems mentioned, is an epistemic virtue, and in this chapter, we shall make a step further and propose and defend the even stronger claim about its central epistemic role; it will be one of the central claims of the book.[4] In later chapters we shall also argue that curiosity, in particular its intrinsic variety, is the foundational epistemic virtue since it is the value-bestowing epistemic virtue. We shall propose a response-dependentist framework, according to which a cognitive state is epistemically valuable if a normally or ideally curious or inquisitive cognizer would be motivated to reach it. But in this chapter, we stay with curiosity's role in motivating and organizing our capabilities and propensities.

Here is the sketch of the route we shall be taking in this chapter. If curiosity is a virtue, as we argued in the previous chapter, it is epistemically a very important one, since it has impressive motivational power. We shall first compare it to other virtues. Some of them are mainly abilities, like intelligence and the like. They don't by themselves motivate their owners to inquire. Others, closer to their moral counterparts, like intellectual humility and courage, motivate a particular, virtuous behavior in the situation of inquiry (and in situations of passive reception of information). Take a brave investigative journalist (I borrow the example from Baehr (2011), Chap. 9 on intellectual courage), who manifests intellectual courage in her work; we naturally assume, given difficulties and risks of searching in the area, that she was extremely inquisitive, and extremely motivated to find out. Or take epistemic humility: a student shows her professor that he has made a stupid mistake arguing for a conclusion (this actually happened to me the day before I am writing these lines). If the professor is intellectually humble, he would gladly accept her correction. Well, the assumption in the humility case is that he would do this because he is interested, at least partly intrinsically, in whether his conclusion is true, and his moves correct; again, a motivating force distinct from humility itself is required, and the motivating force has to do with interest in how things are. So, character-related virtues normally demand

[4] Roberts and Wood similarly see curiosity, or "love of knowledge" as they call it, as a presupposition, or "necessary background of the other virtues". "The love of knowledge provides the intellectual motive for exemplifications of such virtues as humility and courage" (2007: 306). We shall stress a different quality of curiosity, its participating in the very nature of intellectual humility and courage, and leave purely moral virtues on one side.

Baehr (2011) and Inan (2012) also take curiosity to be very important. We shall go into more detail concerning this motivating role than they do.

some kind of curiosity, in order to start functioning at all. (In the epistemically non-central cases, those of almost completely passive reception of information, they demand for their exercise at least caring for truth and for avoidance of error). Now, curiosity offers this motivational underpinning, providing the epistemic goal for virtues-abilities and the opportunities needed by character virtues. Therefore, if curiosity is a virtue, it is motivationally basic. Further, within virtue framework, items that are motivationally basic are basic tout-court, in contrast to merely executive abilities, and to items, like character virtues, that do motivate but only within a framework offered by the basic motive. So, if curiosity is a virtue it is a basic one. Let me call the account of curiosity based on this idea "the motivating virtue account".

We have to address the doubts about curiosity being a virtue in more detail. Here is the plan. We started by noting that curiosity is often good but is sometimes bad. Plutarch talks about meddlesome curiosity and sees malice as the primary motive for it. For example, Roberts and Wood talk about a nameless vicious concern to know:

> Non-instrumental desires for knowledge can also be vicious or immature, through lack of circumspection. Circumspection is a concern for the values that the knowledge, or the pursuit of it, would affect, along with good judgment about the possible effects of the research or the resulting knowledge.
>
> Nazi medical researchers using concentration-camp prisoners as subjects in experiments with obvious and horrendous harmful consequences for the subjects provide clear examples of extreme vice. (2007: 174)

How can then curiosity be a virtue? Fortunately, curiosity is not alone in presenting this problem: many typical moral virtues raise the same kind of issue. Philippa Foot, whose "Virtues and Vices" paper is the *locus classicus* of the debate, discusses the examples of courage, temperance and charity: what about the presumed courage of a villain, about the temperance helping an evil person in his actions, and what about misplaced charity? And she concludes that they are all virtues. Let us consider her strategy and try to apply it to our topic (thanks go again to Duje Kovacevic, with whom I have discussed Foot's ideas). Her main example comes toward the end of her famous "Virtues and Vices" paper from 1978. It concerns what she characterizes as a "sordid murder" done for gain or to get an inconvenient person out of the way, and "done in alarming circumstances or in the face of real danger". She then puts the crucial question: "Should we be happy to say that such an action was an act of courage or a courageous act?". She repeats that the murderer acted boldly, or with intrepidity, and reiterates her question. She notes that "some people insist that they are ready to say" that the act was courageous. "What are we to say about this difficult matter?" she asks again.

Here, she continues, "it might seem that an act displaying one virtue was nevertheless contrary to another. In giving this last description I am thinking not of two virtues with competing claims, as if what were required by justice could nevertheless be demanded by charity, or something of that kind, but rather of the possibility that a virtue such as courage or temperance or industry which overcomes a special temptation, might be displayed in an act of folly or villainy (ibid.). And she comes with the first possible answer, which she attributes to Aquinas. "Aquinas, in his definition of virtue, said that virtues can produce only good actions, and that they are dispositions of which no one can make bad use'. For Aquinas, the murder in the example is not a courageous deed.

She then notes: "T/he common opinion nowadays is, however, quite different" (ibid.). People are ready to say that the murder was done with courage. She then introduces some reservations: maybe people project, and "like to move over to a murder for the sake of conscience" (15); the difficulty stays with us, she suggests.

She notes: "O/ne way out of this difficulty might be to say that the man who is ready to pursue bad ends does indeed have courage, and shows courage in his action, but that in him courage is not a virtue" (15). She has reservations about the solution: "On the strength of an individual bad action we can hardly say that in him courage is not a virtue (ibid.). But she has some weightier reasons in favor or the solution:

> Nevertheless there is something to be said even about the individual action to distinguish it from one that would readily be called an act of courage or a courageous act. Perhaps the following analogy may help us to see what it

> is. We might think of words such as "courage" as naming characteristics of human beings in respect of a certain power, as words such as "poison" and "solvent" and "corrosive" so name the properties of physical things. The power to which virtue-words are so related is the power of producing good action, and good desires. But just as poisons, solvents and corrosives do not always operate characteristically, so it could be with virtues. If P (say arsenic) is a poison it does not follow that P acts as a poison wherever it is found. It is quite natural to say on occasion "P does not act as a poison here" though P is a poison and it is P that is acting here. Similarly courage is not operating as a virtue when the murderer turns his courage, which is a virtue, to bad ends. (16)

Interestingly, in some places Foot seems to return to Aquinas' view that courage for bad ends simply is not courage. Here is John Hacker-Wright summarizing her view in a later paper, "Von Wright on Virtue". Foot adds that we may be willing to describe an action as an "act of courage" in cases where

> the evil end is distant from the action concerned, as when, for instance, a man does something to save his own life or that of his companions in the course of some wicked enterprise, such as an unjust war. (MD 114)
>
> She believes this implies that courage is more than a simple mastery of fear; instead, it is mastery of fear in the context of the pursuit of good ends. Therefore, courage overlaps with daring and boldness, but since the latter conditions may exist in the absence of any commitment to goodness, they are not identical with the virtue of courage. We may say, instead, that the mastery of fear is a necessary, but not a sufficient, condition for the possession of courage. (*Stanford Encyclopedia*, entry "Philippa Foot")

The virtue theorist thus has two broad ways to go. The way one, endorsed by Foot herself, is strictly Aristotelian: only the proper exercise counts as instance of virtue, the improper ones are not really exercises of a virtue. The courage of a whistle-blower is courage, the bravery of an SS officer is not.[5] The same holds for "curiosity" of Mengele and his likes mentioned by Roberts and Wood.

Let us apply this to curiosity and develop Foot's proposal a bit. On one approach, the nosiness criticized by Plutarch is not really curiosity, at best it is pseudo-curiosity, our curiosity. Curiosity, when virtue, call it curiosity +, includes knowledge of appropriateness, and motivation for appropriate exercise. Curiosity–the vicious inquisitiveness, is not really curiosity. Along the same lines we can designate the *thymos* of a whistle-blower courage+, and brand the pseudo-courage of the SS officer as courage–. We would then in general have two sub-species of cognitive intrinsic desire to know, intrinsic curiosity +, and curiosity–, the bad intrinsic curiosity. The first is truly a virtue, the second is not. We can extend it to extrinsically motivated desire and apply the +/– notation analogously. It is all philosophically fine but is not in step with the ordinary usage. People ordinarily talk about the courage of an SS officer or an Islamist kamikaze, and about curiosity's bad exercise. We thus have way two, the ordinary one, where we go along with common usage and the majority, and count good and bad curiosity as genuine items. Then, the value of curiosity is not going to be always positive; the most we can hope for is that it is either (a) default positive but overridable (= prima facie) or (b) mostly positive, and is thus still a virtue (ordinary courage, or loyalty are in the same boat, and the options are applicable to them as well).

I shall leave both strategies in play, the strict Aristotelian, and the ordinary usage one. Consider now the two sub-options of the latter, namely the *prima-facie* one, our (a) and the constant minimal goodness, our (b). The defenders of prima facie moral accounts in general (from British Intuitionists like Ross to contemporaries like Dancy) have the intuition that in the bad case nothing is left from the candidate virtuosity or goodness: there is nothing good about Mengele's curiosity or SS officer's bravery; they are just defeated by weightier negative considerations. But consider courage. a courageous Nazi is a bad person, but still better as a person than a cowardly Nazi. In judging his character, the negative overall aspects counterbalance the positive aspect of courage; the whole is very, very bad, but courage is still better than cowardice. The same holds

[5] For a well-written recent overview and defense of this approach see Battaly (2014), Chapters 1–3.

for curiosity. So, if I had to choose, my own preference would be for the constant value, over mere prima facie value. Fortunately, the two conceptions—constant versus prima facie—are not very different: the (a) or prima facie is just the extreme case, where minimal value tends to zero. So, my account can as well function with a prima facie evaluation.

I hope that the proposal solves the greatest difficulty that the idea of curiosity as virtue faces. To stay with ordinary usage, I hope that we can safely claim that curiosity is virtue, in important and philosophically relevant cases, but that in some of its occurrences it does not function as such. Most importantly, it is not a special trait of curiosity, but something it shares with most character virtues. Someone wishing to argue that courage, temperance and other such traits are not virtues will face similar problems of bad cases. If we want to keep the tradition, and stay faithful to commonsense, we should accept that they, as well as courage, are virtues, and should refine our notion of virtue accordingly, following the suggestions of Philippa Foot.

Once we have done this, we can safely return to the idea that curiosity is an important motivating virtue, and even go further than this. To this task we turn in the next chapter, the central chapter of the book.

References

Baehr, Jason. 2011. *The Inquiring Mind: On Intellectual Virtues and Virtue, Epistemology*. Oxford: Oxford University Press.

Battaly, Heather. 2014. Acquiring Epistemic Virtue Emotions, Situations, and Education. In *Naturalizing Epistemic Virtue*, ed. Abrol Fairweather and Owen Flanagan, 175–195. Cambridge University Press.

Inan, Ilhan. 2012. *The Philosophy of Curiosity*. Routledge.

Roberts, Robert C., and W. Jay Wood. 2007. *Intellectual Virtues*. Oxford University Press.

Sosa, Ernest. 2001. For the Love of Truth? In *Virtue Epistemology: Essays on Epistemic Virtue and Responsibility*, ed. Abrol Fairweather and Linda Zagzebski. Oxford University Press.

———. 2007. *A Virtue Epistemology Apt Belief and Reflective Knowledge, Volume I*. Clarendon Press.

5

The Motivating Virtue Account

Introduction

We shall argue that epistemic abilities and achievements can be understood in terms of inquiry that virtuously opens, sustains and sates curiosity. So, here is the preview. The next section briefly discusses Baehr's proposal to characterize epistemic virtues, in order to fully prepare the ground for our central thesis. Section 3 proposes a kind of Master Argument for the centrality of curiosity, concentrating on our cognitive abilities and the way curiosity organizes the manner we exercise them in order to reach the truth. Section 4 turns to the relation of curiosity to the character-directed epistemic virtues, like intellectual courage and humility. The somewhat longer Sect. 5 addresses the arguments that have been put forward against the centrality of curiosity, some in books and papers, and some in actual discussion of the proposed Master Argument. The conclusion summarizes the two goals of the chapter: first, to argue that curiosity organizes or helps to organize abilities and character traits that have to do with cognition, and second, to offer a sketch of a new and systematic view of both relevant abilities and character traits, or of "epistemic virtues" as they are nowadays called.

N. Miščević, *Curiosity as an Epistemic Virtue*, Palgrave Innovations in Philosophy,
https://doi.org/10.1007/978-3-030-57103-0_5

At the face of it, it seems that the desire for knowledge and understanding, our curiosity in the most general sense, is an epistemic virtue. However, in order to be sure, we first need to check two kinds of conditions. In the preceding chapter we have address the fact that there are sub-kinds and cases of curiosity that are morally problematic; typically cases of low-level object curiosity aiming at private and intimate matters of others (nosiness), or cases of any level connected to morally problematic goals or consequences. The connection can be motivational, for example, when a doctor does research on pain in order to help police torture its victims, or may just involve awareness of bad and highly probable consequences, for example, when a doctor does research on pain out of scientific interest, but is aware that her results will be immediately used by the police. So, the defender of curiosity's virtue has to account for such problematic sub-kinds and cases, and show that their existence is no threat to it.

Now, we need a general characterization of epistemic virtue, so that we might test this seeming. We have talked about the topic before, but here we want to exploit a particularly congenial test. Namely, Jason Baehr has, in his 2016 paper, proposed a short and persuasive list of four characteristics of an epistemic virtue; he is interested in the character virtues, and we shall see that curiosity qualifies as one. The title of the paper is "The Four Dimensions of an Intellectual Virtue" and we shall take a look at each of the dimension he identifies.

The first item on his list is motivational dimension: a subject possesses an intellectual virtue only if subject's possession of this virtue is rooted in a "love" of epistemic goods. Well, it seems pretty clear that the paradigmatic cases of curiosity are cases in which the subject wants to acquire knowledge and/or understanding, the most valuable epistemic states. These are cases of intrinsic motivational curiosity that are central of our understanding of curiosity in general; the extrinsic cases are less clearly motivated by curiosity, although the term is used for them as well. Still, even in extrinsic cases the subject is striving for knowledge, so a "love" (of a more pragmatic variety) is still present.

The next item is the affective dimension, the subject taking pleasure in the exercise of the activity typical of the candidate virtue. Curiosity passes this test with flying colors; in the paradigm, intrinsic case this is obvious, in the marginal, extrinsic case, the inquirer typically enjoys finding out since this may satisfy her ulterior motives.

The third item is the competence dimension: there is an activity characteristic of the candidate virtue on the basis of which it "can be distinguished from other virtues". Indeed, the activity of inquiring is the characteristic activity prompted by curiosity.

The fourth requirement is what Baehr calls "rationality": the subject should typically believe of the activity characteristic for the candidate virtue that it is "epistemically useful" in the given context, and should be disposed to recognize correctly such a context. Here, we have to assume that the typical case of curiosity is relatively intelligent curiosity, in contrast to the degenerate case in which the inquirer embarks upon an inquiry that is epistemically not useful (or even epistemically harmful) for her, and can recognize situations that enable inquiry to become or be useful. But this is the case with all well-established virtues (see the task of recognizing when is courage required, when is generosity really going to help, and where it will make the recipient miserably and systematically dependent on the generous dispenser of goods, and so on), so curiosity is in the same boat with them.

Curiosity certainly passes the test. I think Baehr's list is demanding enough. It is, of course, not *ad hoc*, not prepared especially for curiosity. So, we might rest content; the first task has been performed, and curiosity is highly qualified to count as a virtue.

The Centrality of Curiosity

The next task that stands ahead of our account of curiosity as motivating virtue would be to show in detail how various sub-species of curiosity organize various character and competence virtues. Here, we can offer the briefest of sketches.

Consider first virtues-abilities. They are normally used for arriving at truth, possibly at knowledge and understanding. But if arriving at these is the goal, the typical motivation will be interest, possibly extrinsic, or intrinsic, curiosity in the narrow sense. If extrinsic, it will still involve not just goal but the presence of the motivating factor: compare the inquisitive person with an epistemic *acrates*, who recognizes the practical goal but cannot bring himself to become cognitively interested in it. The motivation will cater for deploying abilities: let me take a look, let me think about the matter, let me calculate and so on. Consider next the intrinsic case. There, curiosity is directing and guiding abilities; of course, their deployment will partly depend on the nature of objects, but the motivation will be the curiosity. Virtues-abilities mobilized will be set in motion by it. Distinguish two kinds of curiosity: the ordinary wide, and the narrow involving excellence and skill at asking right questions. The wide will at least set in motion the abilities: let me look, let me think, let me calculate. The narrow will do more; it might prescribe the tasks in detail: how much looking, with what amount of attention, how much thinking, in what time, and with what intensity. When I watch a theater performance, it is to some extent my interest in performance, the curiosity, that directs my vision ("Don't look at the detail of the faces of opera singers, look at the whole scene!"), my hearing ("Careful, you are missing interesting harmonies!") and my thinking about what is going on ("Stop comparing Tannhäuser to characters from Kierkegaard, and follow the opera, instead of indulging in philosopher's fantasizing!").

In the longer run sustained inquisitiveness might do much more. It might push the beginner into engaging in sustained practice: exercising abilities in order to improve them, and into organizing one's abilities into a system. How does one acquire the skill of observation? By practicing, exercising the ability to see, notice and so on, and exercising attention and so on. Jane is interested in coming to understand a given domain. She might, if really motivated, turn to training one's skill in reasoning: organizing and enhancing sub-skills, noticing premises, or starting from a desired conclusion, look at best support it.

Consider now briefly an interested attempt at mediation offered by Sosa in his (2015) book, described by him as "irenic" (2015: 61). He recognizes that there are many important intellectual virtues that are not knowledge-constitutive. But then, he calls them "auxiliary virtues" (2015: 43 ff.) and claims that their virtue must be understood within the framework of virtue reliabilism. What makes them auxiliary virtues is mostly that their exercise enables us to acquire or sustain the complete competence, related to skill, shape and situation that gives people the capacity to answer important questions. In contrast to these auxiliary virtues the competences are constitutive virtues. I completely agree with Sosa on other character-involving virtues, which I see as hybrid, between epistemology and ethics. However, I would describe curiosity (or, at least curiosity+) as the driving epistemic virtue. It is not constitutive of knowledge, the way the practical goal is not constitutive of the means for it. But as the basic motivating epistemic virtue it cannot be merely auxiliary. Ironically, considerations of relevance concern the means of coming to truth, and thus of fulfilling our desire for knowledge. How can instrumentally important virtues be constitutive of knowledge, while the goal determining ones are relegated to the status of auxiliary virtues?

Let us now pass to character virtues. A part of their goodness is moral: epistemic modesty is good the way modesty is, and so on. Another, epistemic part has to do with their role in acquiring knowledge. And here, we immediately notice the coupling of various kinds and aspects of curiosity with specific virtues. Take, for instance, the dimension of depth, and the kind of curiosity that goes in depth in contrast to a superficial one. It is coupled with character virtues like intellectual perseverance, determination and patience that Baehr groups under the title of endurance virtue (2011: 21). It mobilizes intelligence and other competencies needed for understanding. For illustration, consider a fictional character, call her Maya. She starts by being curious about a "star she heard of on Science channel", a very bright heavenly body that is on the verge of exploding. She learns that it is a stellar system, and its name is Eta Carinae. Now, her determination and patience step in: she learns that the main star of Eta is on the way of becoming a supernova; she than learns about supernovas. The endurance group of virtues is obviously being mobilized and helped by her sustained inquisitiveness, the active curiosity that pushes her forward. Her success in learning makes her more curious about astronomy and so on. On the other side, the kind of curiosity that above all prefers avoiding error will be coupled with intellectual caution, honesty, intellectual modesty and humility. It will mobilize epistemic vigilance.

Baehr himself notices that there is a structure to be recognized in such cases. He talks of character virtues that provided "initial motivation", and he lists inquisitiveness, reflectiveness, contemplativeness, curiosity and wonder (2011: 20). I propose that we take curiosity as basic, and recognize that it provides not only initial motivation, but often sustained motivation. Wonder is more like a prelude to curiosity; Einstein[1] talks about "flight from wonder" to point out that search for knowledge might be occasioned by wonder, but should be sustained by something more active.

Arriving at truth is the central, or at least one of the few central goals of cognitive inquiry. The desire to arrive at truth is, consequently, quite important for human epistemic practice, and the pure, non-instrumental desire, paradigmatic curiosity, is the clearest case of it. Both instrumental inquisitiveness and non-instrumental curiosity motivate human inquiry, and the latter does it in a pure and paradigmatic fashion.

Putting together claims that we defended above, we might arrive at a kind of Master Argument for the centrality of curiosity. First we re-assert the claim we just defended:

(a) It is epistemically very important to be intrinsically motivated to acquire knowledge and understanding.

Then, we note that

(b) most of the character virtues apart from curiosity do not motivate such acquisition, but rather control, in epistemically and morally relevant ways, the process of acquisition undertaken (we have offered examples in Sect. 1).

Of course,

(c) virtues-abilities are not motivating in themselves; they help realize the goals we are independently motivated to achieve. And we know that

---

[1] Here is the quote:

> The development of this thought world (Gedankenwelt) is in a certain sense a continuous flight from "wonder." A wonder of such nature I experienced as a child of four or five years, when my father showed me a compass. (Written in 1946 for "*Autobiographical Notes*", p. 9)

(d) the genuine curiosity is the central intrinsically motivating drive for achieving knowledge and understanding.
(e) So, if curiosity is an epistemic virtue, it is the central cognitively motivating virtue.
(f) Now, it seems that a central motivating virtue, if available, would be basic for the whole system of epistemic virtues (and their contraries). If this seeming is correct, we can appeal to the fact that curiosity is an epistemic virtue, and conclude that
(g) curiosity is the central epistemic virtue, the cornerstone of the whole cognitive enterprise.

We defended (a) above.

We suggested examples speaking in favor of (b). Most epistemic character virtues like open-mindedness, intellectual courage, intellectual modesty, leave the issue of motivation intact.

Of course, the theoretician might appeal to motivation in describing and characterizing them, claiming, for example, that a person is open-minded is she is ready to accept unusual assumption about the topic she is interested in. However, this does not make interest-curiosity into part of open-mindedness, but rather implies that it is the condition of the exercise of open-mindedness, in the manner we noted in relation to curiosity—it is very often a condition of epistemic action and achievement.

The claim (c) continues the same line; I find it almost analytic, but will not insist on this. At least to my knowledge nobody argues that virtues-abilities like good perception or intelligence by themselves normally and routinely motivate their owners. After all, competences are means for the end and some virtue should determine the end or goal. Of course, one can image the agent reasoning in the way that would prompt such a motivation. He could, for example, tell himself: "I just learn I have an excellent eye-sight; let me try to read this bunch of almost illegible letters. I am not at all interested in what is written, I just want to re-affirm the excellence of my eyesight." This kind of motivation might exist, but it is marginal and atypical for normal cognitive procedures.

This gives us, by elimination, the claim (d).

On (e) By "cognitively motivating" I mean motivating for and in the course of cognition.

From (a) to (d) it follows that (e). Now, there are some interesting challenge with (f). The first is the problem of omniscience: if we knew everything we would stop being curious; so the real value does not reside in curiosity. There is a lot to be said, but here we might just note that epistemology is made for humans, not for angels or gods. The counterfactual, even impossible situation in which we come to know everything, does not determine the values relevant for the ordinary, terrestrial life.

But what about character traits like intellectual courage, that are motivating in themselves? What is the relation of such epistemic character virtues to curiosity? We mentioned them in the claim (b), but they deserve much more careful treatment. To this task we shall turn in a moment.

Before we do it, let us not that a part of what is expected from a virtue is the control of the extent of one's engagement: a brave person knows when to engage in a conflict, and when better avoid it, and so on. The virtuous curiosity should do the same with the problem of the amount of research and checking in a search of an answer. It would avoid the Pyrrhonian extreme of endless investigating without any hope of finding an answer and the opposite extreme of gullibility, accepting the first solution proposed by any proponent who is the first to do it.

Curiosity and Other Epistemic Character-Virtues

We have been extolling inquisitiveness-curiosity, or passion for knowledge, virtue clearly linked to truth-goal and have taken it as the basic motivating virtue. Virtuously functioning faculties and capacities, like perception and reasoning seem, in this picture, to be the purely executive virtues that "confer aptness for a good work, viz. the consideration of truth", as Aquinas puts it (*Summa Theologica* I./II, q. 57). But what shall we do with epistemically relevant virtues that are clearly connected with the character of the cognizer? For example, what about open-mindedness, intellectual humility, integrity/honesty, and perseverance, epistemic virtues that seem to be character-traits rather than capacities? Several authors, prominently L. Zagzebski (1996) and S. Haack, have noticed that they are quite similar to moral virtues. The former has proceeded to argue the "special case thesis"; that is, that these epistemic virtues are just a sub-species of the moral ones. "Here I can only try to suggest that a truth-centered virtue-based view has means to accommodate them in a natural and intuitive way; the suggestion will have to rely upon a merest sketch I will argue that the intellectual virtues are so similar to the moral virtues in Aristotle's sense of the latter that they ought not to be treated as two different kinds of virtue. Intellectual virtues are, in fact, forms of moral virtue", writes Zagzebski (1996: xiv). And Haack joins in: "And I am not sure but that here the relation of epistemic to ethical appraisal may be as intimate as the special-case thesis maintains; perhaps, at least without an 'otherwise', 'he is a good man but intellectually dishonest' really does have the authentic ring of oxymoron". And she "suggests a friendly reinterpretation of what is most plausible in Clifford's condemnation of 'the habit of credulity', [...], as pointing to the moral importance of intellectual integrity" (ibid.). Let me then take intellectual integrity as the central example of a character-trait epistemic virtue and propose you to consider a religious scholar who stumbles upon a trace that might lead to the discovery that Jesus never existed, and that Scriptures were written as an attempt at manipulation of the reader. The scholar has s intellectual integrity if she acknowledges that there is such a trace, and intellectual honesty if she goes on to investigate it, say, all the way to the bitter, disappointing end.

Haack is right to point to a moral dimension of honesty and integrity, and I find her and Zagzebski's intuitions convincing. But why do we in fact cherish intellectual honesty and integrity? What is the relation of moral appreciation to the epistemic one? In order to find the answer that would explain the intuitions, compare and contrast our imagined scholar to a proof-finding computer that relentlessly explores the paths to a presumed theorem. The similarity between the person and the machine is that they are reliable in producing information. Each might be admirable for it. The main dis-analogy between the human being and the machine seems to me to reside in the fact that one is a better human being by being intellectually honest and truthful. The scholar is not being merely intellectually honest, but also honest tout court. Her epistemic virtue meshes with character virtue and we may admire her simply for her honesty. The truthful person is morally virtuous by very high Kantian standards; indeed, telling truth is the paradigmatic moral duty, registered as such in central Kantian texts. The two machines sadly lack these excellences. But now, it looks like the virtue is admirable for two kinds of reasons: first, for purely epistemic reasons, which have to do with reliable information producing, and second, for wider, clearly moral reasons, making the person in question a good human being, where "good" has distinctively moral overtones. The first kind of reasons is obviously linked to the goal of arriving to truth. The second is obviously linked to moral considerations. Indeed, the link to character that marks the character virtues often has to do with combating particular weaknesses of character: cowardice or timidity, vanity or rashness.[2]

Return now to the quote from Haack, where she suggests that for character traits "the relation of epistemic to ethical appraisal may be as intimate as the special-case thesis maintains", namely the relation of being a sub-species of. Remember her example: "perhaps, at least without an 'otherwise'", "he is a good man but intellectually dishonest", "really does have the authentic ring of oxymoron". Where does the ring of oxymoron come from? On our diagnosis, it is the moral half of the characterization that produces it: intellectual dishonesty is closely linked to the moral one, a firm symptom, if not one of manifestations of it. The other half is that intellectual dishonesty is deleterious for knowledge acquisition, and this is the properly epistemic half. If this holds, then the special-case thesis is false: moral and epistemic appreciations are distinct, in spite of being structurally similar, and sometimes intertwined, like in the case of character-trait epistemic virtues.

Similar considerations can be produced for the other character-trait virtue, truthfulness. It a communicative virtue: a truthful person will tell the truth even if this goes against her self-interest. If we compare and

[2] We might here apply Morton's idea that cognizers need moral virtues and their epistemic counterparts largely because of their susceptibility to specific emotions (2006).

contrast the truthful person to a reliable search engine, say Google, we note that the similarity between the person and the engine is that they are reliable in transmitting information. Again the main dis-analogy resides in the fact that a human being is a better human being by being truthful. So, why do we normally cherish truthfulness? The first kind of reasons is obviously linked to the goal of obtaining true information. The second is obviously linked to moral considerations, having to do with telling truth even when it goes against one's prudential interest. This is then our diagnosis of what is happening in Zagzebski-Haack intuitions: they diagnose one side of what is in fact the mixed or hybrid character of the virtues they consider.

Turn now briefly to intellectual caution and scrupulousness that counteract rashness and sloppiness. On the cognitive-epistemic side they are useful for avoiding errors. On the practical side, however, they are linked to, if not indeed a part of, general caution and scrupulousness, which are more prudential than moral virtues. So each of them has a cognitive and a prudential component. We might surmise that also in some other cases the supporting virtue in question might be linked more to prudential considerations than to the moral ones.[3]

Consider finally three fine examples of the character-trait virtues given by Jason Baehr in his (2006). First example: a field biologist trying to explain a change in the migration patterns of a certain endangered bird species. He has an eye for detail and remains focused and determined in the face of various obstacles. As a result of his "determination and careful and insightful methods of inquiry", he discovers why the birds have altered their course. Second, an investigative reporter is researching a story on corporate crime and begins to uncover evidence indicating that some of the perpetrators are executives in the very corporation that owns

[3] A further comparison with archery might suggest that the hybrid character of excellences might be a more general phenomenon. In archery, the motivating virtue is some kind of love of archery in the sporting case, and the motive in the non-sporting case might be bloodthirstiness or simple need to hunt. The primary executive virtue is accuracy, perhaps also the ability to calculate the influence of wind and similar potential obstacles. But there are *character-related (supportive) virtues needed in relation to our weaknesses,* like perseverance, concentration and dedication. And they seem to be hybrids again: on the one side we appreciate their specific contribution to archery, on the other, they are generally positive character traits. If this holds, our epistemic story is not *ad hoc*, but points to a wide-spread phenomenon.

his newspaper. In spite of personal threats, the reporter goes on and after several months of research uncovers and exposes the culprits. Third, a historian who has become famous for her view on a historical matter concerning "founding fathers" runs into materials that blatantly contradict her own previous story. Because she is more interested in believing and writing what is true than she is in receiving the praise of her colleagues and readers, she repudiates her influential account, both privately and in "print". Baehr's is diagnosis is that here reaching the truth is not simply a matter of having good capacities, but rather of having certain inner attitudes or character traits. It seems to me that both Baehr's story and his diagnosis perfectly fit our bipartite view. He notes the refined powers of observation and discrimination of his imaginary biologist, which are purely executive virtues on our account, and also notes "his patient, focused inquiry". The valuable result is that the biologist "discovers why the relevant bird species has altered its migratory course". So, one value certainly has to do with the truth of the explanation. On the other hand, patience and ability to focus seem to be general prudential virtues, that here get a particular, cognitive application. Next, his imagined reporter uncovers a corporate scandal because he is intellectually courageous and autonomous. But being undaunted even by personal threats certainly shows general courage: in this case, intellectual courage is just courage, displayed *a propos* an intellectual issue. Finally, his historian acknowledges the error in her work because of her intellectual openness, humility and general love of truth. The last is our central virtue by other name, whereas openness and humility are Janus-faced: on the one hand, they are just particular instances of the homonymous moral virtues, on the other, they play a specifically epistemic role, since they support the quest for knowledge. The enterprise of inquiry is a vast one, and it needs various positive qualities, many of which are only indirectly and distantly connected with the main goal.[4]

[4] An analogy might help. Take any huge enterprise, say, war. The central goal of waging a war is normally victory. It is normally connected to wider goals, either defensive or aggressive, like spreading one's territory. But not any victory is automatically desirable. Pyrrhic victories, if they deserve the name of victory, are hardly desirable: if your army is so exhausted by the victorious war, that it will fall victim to the first insurrection, the victory is better not to be had. What is wanted are victories that are cheap, and total, then victories in important battles, and so on. Victory is *the generic goal*. Consider now the qualities needed for waging war. On the one hand we have narrow military

Let me generalize, and propose a separatist hypothesis concerning character-trait virtues like open-mindedness, intellectual humility, honesty, and perseverance: most of these character trait virtues are to be understood partly in terms of the narrowly epistemic goal(s), in relation to which they play a role of supporting their owner's cognitive efforts, partly in terms of general moral and prudential character evaluation. In this sense, they are hybrid virtues. The separatist hypothesis would nicely account for the intuitions on the truth-focused side, for instance, Sosa's, that the discussion of such virtues falls within ethics of belief and not within epistemology (2015: 43 ff.). I don't endorse these intuitions, but I see their point: the moral aspect of supporting virtues really has to do with ethics, rather than epistemology, whereas their cognitive aspect is squarely within the domain of epistemological concerns. The present proposal balances two contrasting intuitions, the one of character epistemologists, who would focus epistemology on such virtues, and the Sosa camp, that would completely exclude them from epistemology. It suggests that each is almost right about one "half" of hybrid virtue, and wrong about the other half.[5] So much about the separatist hypothesis

abilities, that concern fighting itself. They come in two varieties, the motivating virtues, like military zest, perhaps also patriotism, then courage, steadfastness, authoritativeness of character for officers, and the like, and purely executive virtues, like physical endurance, strength, being a good marksman, and the like. But winning a modern war is also matter of having all sorts of support, prominently logistic and medical support, but also support of civilian population. A good contemporary army general will be very keen on this background support. Finally, a good military doctor is good military doctor only if he is a good doctor, period, plus if he is particularly adroit at healing, say, wounds, in contrast to being good at plastic surgery. Similarly, cognitive enterprise involves a whole lot of virtues that are only tangentially or indirectly connected to the main goal.

[5] Let me mention two challenges to the idea that character-related traits are "merely" supportive. First, a general challenge. Montmarquet (1993, 2000), maintains that the traits in question are intellectual virtues on account of certain of their internal or psychological features considered *in their own right*. This account has the advantage of being able to explain the apparent *personal* worth or value associated with these traits. This is difficult to account for on a model of intellectual virtue (like Driver's) according to which the traits in question are strictly instrumentally valuable. Answer: Montmarquet's characterization of internal considerations and features being valuable when considered in their own-right is purely moral; it is thus covered by the moral aspect of a hybrid value.

Second, a particular challenge from moral knowledge & self-knowledge. Robert C. Roberts and W. Jay Wood argue that epistemic virtues are character virtues, "intimately connected" with person's emotion. But the example they give have to do with reacting to a scene involving a racial injustice, hardly recognizable by someone not emotionally tuned in.

(web: 9). The example clearly involves emotional sensitivity. But it is a special case, where morality is the topic of the debate and of cognition. If we assume that emotional sensibility is part of the

claiming the hybrid nature of character-trait epistemic virtues. Here I cannot go on to try to prove it on the case by case basis, since this would require a separate paper at least. I will keep it as the best guess, which might help us to organize and regiment the swarm of epistemic virtues.

Finally, character-trait epistemic virtues are sometimes being linked to goals apparently different from truth; for instance, W. Riggs (2009) links open-mindedness and insightfulness to understanding, rather than to truth. He might be right, and we have to make our schema a bit more complicated. The supporting virtues would still be, but their cognitive component of hybrid virtues probably has multiple links: not only to truth (and knowledge), but also to understanding, and other epistemic values. The value of truth is just the general, common denominator of all others. Where does it come from? What is its ground? We shall turn to this issue in Chap. 7. Here, we have a more urgent task: address the objections that have been put to the idea of strong, curiosity-centered virtue epistemology.

---

capacity to recognize the moral significance of situations, and I agree with the authors that this is the right move, emotional sensibility will be seen a cognitive (sub-)faculty for the special domain of moral cognition, and there is no wonder it would plays a central role there. This does not show that executive epistemic virtues are identical to moral ones, nor does it in itself tell anything about the hybrid view of character-related epistemic virtues like intellectual humility.

The critics might go on to argue that character-related epistemic virtues generally involve the same kind of emotional component, inseparable from their cognitive role, in contrast to the separatist view defended here. An intellectually humble person recognizes emotionally when it is time to retreat, they might continue. And will thus learn more from genuine philosophical authorities, say, from Putnam or Lewis, than the intellectually arrogant one: the meek and the humble shall inherit the Twin Earth, as the saying goes.

Answer: epistemic humility involves retreating for epistemic reasons, recognizing the superior intellectual standing of one's superior. If I simply shut up in a discussion out of moral humility, feeling that continuing the debate would be seen as disrespectful, although it seems to me that I am theoretically right, I don't exercise an intellectual virtue, but a purely moral one. I might thereby be also exercising an intellectual vice. On the other hand, intellectual humility is cognitively useful and epistemically justified, pointing (fallibly) at the same to a generally non-arrogant character. It thus has epistemic value, and can be of moral interest as well, exactly as the separatist view would predict.

References

Adam, Morton. 2006. Imagination and misimagination, In Shaun Nichols (ed.), The Architecture of the Imagination: New Essays on Pretence, Possibility, and Fiction. Clarendon Press.

Baehr, Jason. 2006. *Character, reliaility and virtue epistemology*, Philosophical quarterls, 56(223): 193–212.

———. 2011. *The Inquiring Mind: On Intellectual Virtues and Virtue, Epistemology*. Oxford: Oxford University Press.

Linda, Zagzebski. 1996. *Virtues of the Mind: An Inquiry Into the Nature of Virtue and the Ethical Foundations of Knowledge*, Cambridge University Press.

Montmarquet, J. 1993. *Epistemic Virtue and Doxastic Responsibility*, Lanham, MD: Rowman and Littlefield.

———. 2000. "An Internalist Conception of Epistemic Virtue," in G. Axtell (ed.) Knowledge, Belief, and Character, Lanham, MD: Rowman and Littlefield.

Riggs, Wayne D. 2009. Understanding, Knowledge, and the Meno Requirement. In *Epistemic Value*, ed. Adrian Haddock, Alan Millar, and Duncan Pritchard. Oxford: Oxford University Press.

Sosa, Ernest. 2015. *Judgment and Agency*. Oxford University Press.

6

Defending the Motivating Virtue Account

Introduction

Here is then the plan. In Chap. 4 we looked at the prima facie impression that curiosity, in the wide meaning of desire for knowledge and understanding, is virtuous, and then addressed two kinds of doubts. We concluded by reaffirming the impression: the doubts are exaggerated, curiosity behaves like other, well-established virtues and we may count it as one. In Chap. 5 we developed an even more curiosity-friendly approach, our motivating virtue account, and argued for the centrality of curiosity as the crucial motivating virtue; the account offers, almost as a by-product, a systematization of epistemic virtues that might prove useful.

Here, we turn to further objections against the privileged epistemic status of curiosity. An important objection is answered.

Even the virtue turn, in recent epistemology, did not make curiosity very popular. Curiosity seems to be something of a character trait; but character-based virtue epistemology did not concentrate much upon it.[1] The proponents of virtues-as-abilities wrote little about it, and when they

[1] For a valuable exception see Roberts and Wood (2007). Roberts and Wood see curiosity, or "love of knowledge" as they call it, as a presupposition, or "necessary background of the other virtues".

N. Miščević, *Curiosity as an Epistemic Virtue*, Palgrave Innovations in Philosophy,
https://doi.org/10.1007/978-3-030-57103-0_6

did, they were more critical about it than full of praise; E. Sosa is the prime example, and here I shall further develop my own proposal in dialogue with his views (with thanks to Ernie for the actual questions and conversations!). The other author of an important objection is M. Brady. However, the normative part is still largely to be done. In the next section I shall briefly situate the motivating virtue account in relation to some of the main proposals in contemporary scene. And toward the end of the chapter we shall briefly address a possible objection, having to do with the aretaic status of curiosity: can it be a virtue if it is dependent on emotional states of the agent?

Criticisms and Replies

The stress on virtues is typical for character-stressing—responsibilist thinkers, like Zagrebsky and Baehr. In spite of differences between them, they all agree on the importance of the motivating role, and in this respect are close to our account. However, our account is open to the possibility that curiosity does the central motivating work, but that the achievements entering the definition of knowledge belong to the competences-abilities. This is in fact the solution I prefer, but will not argue for it here; as far as this book is concerned, I am satisfied with the motivating role of curiosity and I leave open the issue of the carriers of the constitutive-definitory role.

On the opposite end are competence-stressing views. The simple version has been proposed by Greco and Sosa; the latter defends it in his earlier work, including the 2001 paper, that we shall discuss in the moment. Here is a typical formulation:

> When a belief is correct attributably to a competence exercised in its appropriate conditions, it counts as apt and as knowledge of a sort, animal knowledge. (2007: 92)

The love of knowledge "provides the intellectual motive for exemplifications of such virtues as humility and courage", they write (2007: 306).

The more sophisticated is the responsibilist competence view, put forward by Sosa, in his 2015 book. Here, the responsibilist aspect becomes prominent in the constitution of reflective aspect of knowledge.

Our motivating virtue account divides the roles: the character virtue of curiosity is responsible for motivating and organizing virtues-competences, and the latter are responsible for achieving knowledge. (Two authors that have recently come to a very similar account are Fairweather and Carlos (2018), and Robert T. Pennock has recently offered a congenial account of scientific curiosity in his 2019 book.)

Those who are besmirching curiosity have a battery of arguments that fall into three categories. The first, well represented by Sosa's remarks (2001) concentrates upon the cases of idle interest, the second, upon volatility, irrelevance, superficiality and similar defects: this one is exemplified in the earlier period by Burke's remarks, and defended these days by Michael Brady, who relies on the work of the psychologist Paul Silvia (2006). Here is the quote from Burke:

> Curiosity is the most superficial of all the affections; it changes its objects perpetually; it has an appetite which is sharp, but very easily satisfied; and it has always an appearance of giddiness, restlessness and anxiety. (Edmund Burke, *A Philosophical Enquiry into the Origin of Our Ideas of the Sublime and Beautiful*)

Finally, there is an objection that targets the idea of epistemic high usefulness of curiosity, call it "Redundancy Objection", again due to Sosa (and Michael Watkins): excellent epistemic results can be obtained without intrinsic, paradigmatic curiosity, so what is epistemically so special about it? It comes in several varieties that we shall address in turn.

The Idle Curiosity Objection

Let us start with the first, the Idle Curiosity Objection, as formulated by Sosa:

> Consider a randomly selected cubic foot of the Sahara. Here is a trove of facts, of the form grain x is so many millimeters in direction D from grain y, than which few can be of less interest. Or take some bit of trivia known to me at the moment [...] I confess that I will not rue my loss of this information, nor do I care either that or how early it will be gone. As interpreted so far, the view that we rationally want truth as such reduces to absurdity, or is at best problematic. (2001: 49)

There several ways to answer the objection. One answer is to claim that curiosity about trivia is not a virtue, not a curiosity+; the virtuously curious+ person knows what intellectual interest to pursue, and will not easily get into counting grains of sand. One can go further and note that some pursuits are trivial in one context and worthwhile in others. The history of chewing gum either is or would be a worthwhile part of cultural studies: there is a fine documentary about Thomas Adam who invented it almost by accident, and Bill Wrigley who made a fortune out of it. What about Sosa's example of the interest in the number of blades of grass? In itself it is just eccentric; what makes it non-virtuous is that its point is normally overridden by other interests. Suppose the grass counter asks Sosa what else she should do. One possible reply is that there are prudentially and morally more important things and there are also cognitive alternatives to counting grass: read a good book that will teach you about something important, or sit down and think about the Gettier problem (although, for most people outside philosophy circles, this is, unfortunately, almost indistinguishable in usefulness from counting blades of grass). The point is that *the only epistemic alternative to being curious in one thing is being curious in another*: the contrast between our grass counter and a deeply thinking philosopher is not the contrast between curiosity and something else, but contrast within the family, between people being curious about different matters and wondering about them. What we do is weigh the virtues against each other and do the same for values.

There is a second way to answer, and it comes with two sub-options. Here is the first one. There is something *prima facie* good about every instance of curiosity, at least epistemically good, but this goodness is overridable (or merely present *pro tanto*, depending on which theory of value-properties you believe in): interest in the data on grains of sand is ok in itself, but normally it is overridden by more urgent and prudentially/morally more important ones. This is my favorite option.

However, if you are more of a particularist, you can take it that it loses its value completely, once it is confronted by the stronger competitor. Let me just mention an example in which the interest in the number of grains of sand has not been overridden or defeated. See the following piece of news:[2]

> It might sound like a never-ending job, but scientists have started counting grains of sand on a Cornish beach to try and calculate the rate of coastal erosion. And Plymouth University researchers reckon it will take about five years to carry out their study, monitoring data from instruments mounted on 40 metres of scaffolding at Perranporth. Data will be collected using a bank of laptop computers housed in a mobile field laboratory installed at the top of the beach.

So, the interest in the number of the grains of sand is not always misplaced. The critic might still feel that the interest is in itself misplaced, but this feeling must then rest on some implicit conception of what epistemic value consists in; we shall return to this issue shortly, when we come to the "Redundancy Objection".

Let us now just mention one more sub-option: the quality of the majority (of interests) decides whether the type is good or not. If courage were used mostly in a nasty way, say, in a Nazi world, it would stop being a virtue. What is the verdict for curiosity? Well, the range of our interest is very wide. Of course there will be some irrelevant ones around, but the majority is relevant, and curiosity is, one the one hand, necessary for survival and flourishing life, and on the other an almost defining feature of our cognitive lives.

[2] Available here: http://www.dailymail.co.uk/sciencetech/article-2048563/Every-grain-counts-The-scientists-literally-counting-sand-erosion-study.html#ixzz2zeK3LndC.

Superficiality Objection

Our next topic is the Superficiality Objection; we shall look at the recent version proposed by Michael Brady. In a handout available on his web page, titled "Curiosity & Epistemic Goodness" he writes: "There is wide agreement—among psychologists, at least—on the appraisal variables that generate interest." He also writes, "One of the central appraisals is of novelty: 'whether or not an event is new, sudden, or unfamiliar. For interest, this novelty check includes whether people judge something as new, ambiguous, complex, obscure, uncertain, mysterious, contradictory, unexpected, or otherwise not understood'" (Silvia, *Exploring the Psychology of Interest*, 2006, 57). He also mentioned that interest and importance diverge. In his paper on "Curiosity and the Value of Truth" he pointed out that we tend to find old, expected, familiar things comfortable or enjoyable, but are interested in things which are unexpected, unfamiliar, mysterious, baffling. Brady also mention a related puzzle (in his 2009 paper "Curiosity and the Value of Truth", in Adrian Haddock, Alan Millar & Duncan Pritchard (eds.), *Epistemic Value*. Oxford University Pres): there might be trivial questions in relation to which there is no point in being curious, but that might have surprising and worthwhile answers, and exciting questions which have completely trivial and uninteresting answers. So, curiosity is not correlated with the value of its object: sometimes we are rightly uninterested in a question that has interested answer, sometimes we are rightly obsessed by the question which has uninterested answer. I surmise that the answer lies in the focus of curiosity.

Let us answer starting from the possibility he appeals to. Let me introduce, for illustration, an imagined variant of Othello, call him Thello. Suppose that Thello completely lacks curiosity concerning the whereabouts of a handkerchief that he long time ago offered to his beloved Eudemona; however, if he were to inquire, he would found his present in the hands of Eudemona's secret lover. Of course, in order for the example to work, Thello must be very interested in whether Eudemona is faithful; otherwise the final result would be as boring to him as the initial question.

Inversely, Othello can become terribly interested in the handkerchief of Desdemona, thinking that it conceals a terrible secret, whereas the true answer is, unexpectedly for him, completely trivial. In general, we are curious about more than one topic at a given time, and therefore questions that don't interest us from the viewpoint of their explicit topic (e.g. handkerchief) can generate answers that interest us a lot from the viewpoint of topics they themselves introduce (e.g. my darling having a secret lover). Inversely, Othello can become terribly interested in the handkerchief of Desdemona, thinking that it conceals a terrible secret, whereas the true answer is, unexpectedly for him, completely trivial. Nothing in our stories tells anything against relevance of curiosity.

So, the curious person starts be noting that something is ambiguous (complex, obscure, mysterious, contradictory), and asks oneself how one should one understand it. To me it seems that *if curiosity is directed to the* "new, ambiguous, complex, obscure, uncertain, mysterious, contradictory, unexpected, or otherwise not understood", then its central goal is achieving understanding, rather than arriving at isolated items of knowledge, and I think it is epistemically quite a good thing. The interest in complexity leads to the desire to understand, the crucial epistemic desire. Novelty is in the vicinity; it involves not-yet-understood matters. Finally, a virtuous researcher is able to control herself, to balance novelty with relevance and depth, and so on. So much for the first line of defense. But one may also add that the interest in the novel and the complex is, globally seen, extremely epistemically useful. The novelty liberates us from cognitive inertia; just think of depressed people who have lost their natural curiosity. Finally, not all news from cognitive science are pessimistic. Here two well-known psychologists offer a more optimistic view:

> Curiosity, novelty-seeking, and openness to experience are all associated with desirable psychosocial outcomes. This includes general positive affect, willingness to challenge stereotypes, creativity, preference for challenge in work and play, perceived control, and negative relationships with perceived stress and boredom [...]. The emotional-motivational state of curiosity appears to fuel positive emotions such as excitement, enjoyment, and attentiveness [...], facilitating complex decision making [...] and goal perseverance. (Peterson and Seligman 2004: 134)

The Redundancy Objection

Finally, we have the Redundancy Objection. The first variant is the following. In his more recent (2015) book, Sosa stresses that epistemology is not a department of ethics. He notes that an extremely high epistemic status, certain knowledge, can be attained with a deplorable state that represents a sad waste of time. His example here is someone spending a morning determining with certitude how many beans are left in their coffee bag. At the Taipei conference Sosa kindly asked me a question that was another version of this objection: we achieve good epistemic results out of all kinds of motivation, extrinsic as well as intrinsic. In his *Judgment and Agency* he develops the same approach:

> [T]here is a distinctive dimension of epistemic assessment isolated from all such broadly ethical (or prudential) concerns. Moreover, within this epistemic dimension, love of truth plays a negligible role *at most*, if any at all. Hedge fund managers, waste disposal engineers, dentists, and their receptionists, can all attain much knowledge in the course of an ordinary workday despite the fact that they seek the truths relevant to their work only for their instrumental value. That is why they want them, not because they *love* truth. (2015: 48)

A very similar point was made by Michael Watkins (in the discussion at the Rijeka conference in 2014). He compared an incompetent but genuinely intrinsically curious person with a non-interested (or merely extrinsically motivated) professional: the former never arrives at knowledge, and the latter acquires it without any problem. So, curiosity (in its genuine, intrinsic form) seems to be redundant. Let me try to answer first Sosa and then Watkins.

Let me concede to Sosa that no character virtue (certainly including motivating ones) is constitutive for knowledge. I agree. I would prefer to describe curiosity as the driving epistemic virtue. It is not constitutive of knowledge, the same way the practical goal is not constitutive of the means for it. But what is more important: a constitutive, essential element of knowledge, or the source organizing the whole epistemic field? Note that we need other items in epistemology besides knowledge: true belief and understanding, for instance. Curiosity enables us to attain knowledge (and understanding, and so on.). But are all enabling conditions ipso *facto* merely auxiliary? Some enabling conditions go beyond merely auxiliary. For example, a global enabling condition might be something like "transcendental condition" of existence. But then, non-constitutivity is a red herring.

Sosa's no-constitutiveness point does contain a deep insight, and the defender of the motivating epistemic virtue has to take it into account. What the defender of the motivating epistemic virtue has to do is to widen the notion of curiosity. This is not an *ad hoc* move since this is being done in psychology. I mentioned that the standard work by R. Sylvia speaks of "interest", a motivational state encompassing various species. so we may talk of "curiosity" in wide sense, divisible into intrinsic versus extrinsic, instrumental motivation; I did this already. Instrumental-extrinsic curiosity still involves not just the goal but the presence of the motivating factor:

Consider first the instrumental case. Compare the inquisitive person with an epistemic *acrates*, who recognizes the practical goal but cannot bring himself to become cognitively interested in it. The motivation will cater for deploying abilities: let me take a look, let me think about the matter, let me calculate and so on. Consider next the intrinsic case. There, curiosity is directing and guiding abilities; of course their deployment will partly depend on the nature of objects, but the motivation will be the curiosity.

In practice, the pure types are less common than the intermediate ones, combining some pure, intrinsic motivation with an instrumental kind. The inquirer starts from the instrumental, but then becomes interested for the sake of solving the problem (functional autonomy of the motive)

One is primarily intrinsically motivated, but there is an accompanying instrumental motivation (e.g. Both Darwin and Hawking talk about their curiosity and the competitive drive as motivating them. The successful inquirer finds the balance: Pasteur wanted at the same time to do serious science and help humanity; he found the area of medicine when one can do both.

So, the first version of Redundancy Objection seems to cut both ways: even not very intelligent curiosity can lead one to achieve high-status knowledge, and this tells something in favor of curiosity. If it's most degenerate forms can and do lead to states with "extremely high epistemic status", doesn't this speak in its favor, rather than against it? The independence from considerations of ethics enhances the point: if we consider the counting of beans purely epistemically, and ask for its epistemic value, and we take into account that the result has high epistemic status, then the purely epistemic value of curiosity about beans is not zero, and not negative; it is positive, although perhaps not particularly impressive.

The version offered by Watkins contains a comparative element. Let me resort to analogy. So, in order to see that the question is not completely fair, imagine a similar objection being raised against the importance of courage. A skinny, unarmed and not very strong teenage hooligan is attacking passersby with his fists. Big John is not a particularly courageous person, but he is twice the size of the hooligan, and endowed with strong muscles, whereas the other passer-by, Small Bill, is a small, physically very week but quite courageous person. Which of the two will have fewer problems with the hooligan? Big John, of course; the hooligan will give up at the first sight of him; so, courage is redundant.

In short, the example is unfair, pitting courage-coupled-with-weakness against physical strength coupled with very little courage. The virtue loses. The curious amateur has no chance; this does not show that curiosity is generally redundant, the same way in which the street fight example does not show that courage is irrelevant.

However, the crucial consideration in the debate is the thesis that epistemology is independent of ethics. Definitions of knowledge look at excellence of cognitive achievement, not to its possible motives, the objector might insist. In this respect cognition is like archery or art of dancing: what counts intrinsically is the skill and the shot in the first place, the skill in performance in the second. Some do it for money, other for fun, a third group might do it from extremely bad motives (shooting an innocent victim), but what counts in the judgment of skill are not these external motives.

I would reply that in the cognitive case there is a contrast between epistemic and non-epistemic motivation, which cannot be irrelevant to epistemology. It does not enter the definition of knowledge but it can still be constitutive for the general form of epistemology, in particular virtue epistemology. With genuine, intrinsic curiosity, we have *epistemic* motives that are *internal* to the cognitive activity. We should not abstract from them when deciding about the shape of our epistemology. It should be made for humans as they are, endowed with such epistemically internal motivation. To see this most clearly, imagine a close but distinct species, that has only practical interests and no intrinsic curiosity (this being the only difference between us and them); the members of the species can achieve good cognitive results if this brings them some profit, but they are never, never interested in how things really are, beyond the purely practical interest in applicable knowledge. Our species might be moving in this direction, at least judging by the ways research project are being financed, but for the time being, the cognitive life of members of our imagined species is quite different from human cognitive behavior. Here is a quote typical for a great scientist:

> My colleagues in elementary particle theory in many lands [and I] are driven by the usual insatiable curiosity of the scientist, and our work is a delightful game. I am frequently astonished that it so often results in correct predictions of experimental results. (Murray Gell-Mann, Nobel Banquet Speech (10 Dec 1969), in Wilhelm Odelberg (ed.), *Les Prix Nobel en 1969* (1970))

For humans it is epistemically very important to be intrinsically motivated to acquire knowledge and understanding. This does not depend on ethics; curiosity is an epistemic character virtue, but it is crucial for the character of a genuinely impressive cognizer.

The objector might at this point push the comparison with dancing: it is the skill, not the motivation that decides. And achieving knowledge is in this respect like dancing. Answer: yes, the motivation is irrelevant when you abstract from the whole of the performance. But you should not; to see this suppose that the dancer is at the same time the choreographer. And she wants, at a given point, to present an ordinary walk as an artistic performance. Then the quality of the walk-as-a-piece-of-dance will depend to a large extent on the overall artistic motivation. And this projects to the case in which dancer and choreographer are distinct persons. These matters are nowadays routinely discussed in connection with dance-like performances, and aesthetic, intrinsic motivation (actual or ideally reconstructed) counts there as an element in the quality of the work.

Virtue and Emotion

So much about actual objections. Let me conclude with a possible one. I am treating curiosity as epistemic virtue; however many authors connect it strongly with emotion; in the chapter on cognitive psychology we shall encounter the example of Paul Silvia who takes it as obvious that curiosity is "an emotion", as he puts it. A virtue epistemologist or ethicist might object that emotions are too unstable to constitute a virtue.

Let me briefly address this objection, since the topic is a prominent one in the recent discussion. In general, important authors are linking emotions and virtues, in particular epistemic ones. The connection between morality and emotion is, of course, a long-lasting topic in moral psychology. Let me remind the reader of a famous encounter: Jürgen Habermas uses Strawson's analysis of sentiment as the foundation of his own theory of moral justification of action. He is very decisive about central function of emotion in moral justification. He writes, "Feelings seem to have a similar function for the moral justification of action as sense perceptions have for the theoretical justification of facts."[3]

He sees Strawson as his precursor, in particular his comments on resentment. He notes that "Strawson begins by examining an emotional response which in its obtrusiveness is well suited to convince even the most diehard skeptic that moral experience has real content" (45). And he points out that "when there is no restitution for the initial injury, our unambiguous reaction will harden into smoldering resentment. This

[3] Habermas, Jürgen (1990), *Moral Consciousness and Communicative Action*, MIT Press, p. 45 ff.

enduring emotion lays bare the moral dimension hidden in every insult" (ibid.).

Let me quote his comment about the way Strawson makes several "important points" about the connection:

> When actions violate the integrity of another person, the perpetrator or a third person may produce excuses. As soon as an aggrieved party accepts an excuse, he no longer feels injured or slighted in quite the same way as before; his initial indignation will not turn into lasting resentment. Excuses are like repairs we make to disturbed interactions. (1990: 46)

He notes the usual pointing toward the competence of the actor. "The latter is precisely what we do in the second kind of excuse when we point out that the wrongdoing was the act of a child, a drunk, a madman, etc., that it was committed by someone who was beside himself or incapacitated by stress, as from the effects of a serious illness, and so on. This second type of excuse invites us to view *the actor himself* in a different light, to wit, as someone to whom we cannot unqualifiedly ascribe the qualities that characterize a competent subject. In this case we are supposed to take an objectivating attitude that precludes any moral reproach from the start."

"Strawson's third point leads to the moral core of the emotional responses discussed above. Indignation and resentment are directed at a *specific* other person who has violated our integrity. Yet what makes this indignation moral is not the fact that the interaction between two concrete individuals has been disturbed but rather the violation of an underlying *normative expectation* that is valid not only for ego and alter but also for all members of a social group or even, in the case of moral norms in the strict sense, for all competent actors" (47).

Similarly with the movement of connecting emotion with virtue. The movement has been taking inspiration from the Aristotelian tradition. Let me mention three authors from the tradition whom I find particularly helpful: Rosalind Hursthouse (2000), Heather Battaly (2014) and, most recently, Kristján Kristjánsson (2018).

Here is a proposal for characterizing good curiosity as an Aristotelian virtuous emotion, inspired by Kristjánsson's treatment of gratitude (2018:

57).[4] First, virtuous curiosity constitutes a stable emotional trait: the appeal to trait and stability is important for the virtue theory, and the emotional characterization answers our present concern. Second, it is the trait of craving to find out certain things; the specification of "craving" would take us to the psychology of desire, and we gladly leave this for another occasion. The craving and finding out should be formed in a morally sustainable way: for an instance of bad curiosity see the last lines of this paragraph. What the person is trying to find out are facts and connection. The facts alone give ordinary knowledge that connections support understanding. Knowing how things are connected enables one to understand, and craving for such knowledge is a legitimate, I would even argue crucial, piece of virtuous curiosity. The facts and connections are presumably important for one.

Next comes the hedonic element: the craving is to a large extent pleasant. Not always; going to a medical lab to find out if one has been diagnosed cancer is normally accompanied by intense interest in the result, but the interest is hardly pleasant, certainly not to a large extent. Finally, possessing the trait is intrinsically valuable; having the right curiosity is a component, probably and an indispensable one, of a flourishing life.

References

Battaly, Heather. 2014. Acquiring Epistemic Virtue Emotions, Situations, and Education. In *Naturalizing Epistemic Virtue*, ed. Abrol Fairweather and Owen Flanagan, 175–195. Cambridge University Press.

Fairweather, Abrol, and Montemayor Carlos. 2018. Curiosity and Epistemic Achievement. In *The Moral Psychology of Curiosity*, ed. Lani Watson et al., 199–216. Rowman & Littlefield International.

Hursthouse, Rosalind. 2000. *On Virtue Ethics*. Oxford University Press.

Kristjánsson, Kristján. 2018. *Virtuous Emotions*. Oxford University Press.

Peterson, Christopher, and Martin E. Seligman. 2004. *Character Strengths and Virtues a Handbook and Classification*. OUP.

Roberts, Robert C., and W. Jay Wood. 2007. *Intellectual Virtues*. Oxford University Press.

[4] Of course, all the responsibility for possible weaknesses and mistakes goes to the present author.

Silvia, P. J. (2006). *Exploring the Psychology of Interest*. New York: Oxford University Press.

Sosa, Ernest. 2001. For the Love of Truth? In *Virtue Epistemology: Essays on Epistemic Virtue and Responsibility*, ed. Abrol Fairweather and Linda Zagzebski. Oxford University Press.

———. 2007. *A Virtue Epistemology Apt Belief and Reflective Knowledge, Volume I*. Clarendon Press.

———. 2015. *Judgment and Agency*. Oxford University Press.

7

Epistemic Value

Introduction

Many philosophers, scientists and educators agree that knowledge has a value. Ilhan Inan is surely among them: in his brilliant curiosity book (2012) he writes:

> Whether curiosity is taken to be a form of virtue or not, it should be clear that there are important connections between being curious and some of our basic epistemic attitudes and achievements. Knowing, for instance, is an epistemic achievement, at least in certain cases, and curiosity is one of its basic motivators. (2014: 143)

Other, related epistemic items like true belief or understanding also seem to have a value. Some of the value seems clearly instrumental. Knowledge that my two new neighbors are happily married to each other might be useful for me in order to know what to expect from them, how to behave toward each and both, and so on. Knowledge that some Neptune moons are rich in water might turn out to be practically useful in a more distant future, when we might need water from outside our usual earthly sources. But, other kinds of value are also in the offing. I was happy to learn about

N. Miščević, *Curiosity as an Epistemic Virtue*, Palgrave Innovations in Philosophy,
https://doi.org/10.1007/978-3-030-57103-0_7

the Neptune moons not because I expected that I will need some water originating from them, but simply because I saw it as a very interesting fact about our distant neighbors in the solar system. Call this another kind of value "intrinsic". Call the value of epistemic states (or facts) "epistemic value", or "e-value" for short. I shall assume that instrumental e-value is not problematic, and concentrate on the intrinsic e-value.[1]

Allow me a few terminological proposals. Take the basic epistemic item, of your choice: truth, true belief, justified true belief, knowledge and understanding. Let me for now call it just "grasping".[2] I shall abbreviate grasping the truth that *p*, as "G*p*". Let us agree that some items of these kind do have intrinsic e-value. For example, "Some Neptune moons are rich in water" has had such value for me. Consider now some *p* and grasping the truth of it (G*p*). One option concerning their value is that they are not valuable; at a more general, philosophical level then, no such items have epistemic value. We have already embraced a more optimistic view, according to which such items often are valuable. Now, supposed I am interested in "p", and curious whether things are as the proposition *p* represents them to be. Again, there are two options. I might be curious about the items as a means to an end, extrinsically and instrumentaly. Knowledge that my two new neighbors are happily married to each other has often this extrinsic character, having to do with expectations and useful ways to behave. Alternatively, some topic might be intrinsically interesting, epistemically attractive in itself. Following Brady (who in this context also mentions Hurka (2001: 6)), I shall equate intrinsic goodness with non-instrumental goodness, leaving aside complicated cases where one can, in a sense, have one without the other (Brady 2009: 265).

Intrinsic curiosity and intrinsic e-value will be our topic. I have already quoted Inan. Let me add one more quote connecting curiosity to epistemic value:

> Human curiosity may or may not be considered to be a virtue; but even if it isn't, it must still be of vital importance in its relation to certain epistemic

[1] I am leaving aside here the difficulties concerning the instrumental e-value put forward by Allan Hazlett in his 2013 work.

[2] Following Duncan Pritchard's (2014) terminology of grasping the truth (either true belief, or something richer and closer to knowledge).

attitudes that most of us value. We wish to be inquisitive and open-minded, and we wish to realize how fallible our beliefs are and become aware of our own ignorance and our cognitive limits. This requires epistemic self-reflection. But where would one fi nd the motivation to do this? There are a lot of things that we do not know, but only a small portion of them is brought to our consciousness. Why is that? Because we care about certain things and not others. We have an interest in certain topics, and we care to know more about them. It is this kind of interest that motivates us to refl ect on our ignorance, and only then we become curious. So in this sense, curiosity is value laden. We are curious only about things that we are interested to know. Such an interest surely is a product of what we value. Even if we don't value the very object of our curiosity, we are interested in it because we believe that coming to know it relates to certain things that we do value. The broader our interests are, the broader the scope of our curiosity. (2012: 183)

Let me next borrow three more quotes listed by Stephen Grimm in his 2008 paper, to illustrate the fact that epistemologists normally accept that some items have intrinsic e-value:

[Goldman:] Our interest in information has two sources: curiosity and practical concerns. The dinosaur extinction fascinates us, although knowing its cause would have no material impact on our lives. We also seek knowledge for practical reasons, as when we solicit a physician's diagnosis or compare prices at automobile dealer shops. (Goldman 1999: 3; emphasis added)

[Alston:] [Although having true beliefs furthers our practical goals] the attainment of knowledge and understanding are also of intrinsic value. (Alston 2005: 31)

[Lynch:] We care about the truth for more than just the benefits it brings us …. There are times in our lives when we simply want to know for no other reason than the knowing itself.

Curiosity is not always motivated by practical concerns. Consider extremely abstract mathematical conjectures. With regard to at least some

> such conjectures, knowing their truth would get us no closer to anything else we want. (Lynch 2004: 15–16; all three quoted in Grimm 2008: 727)

I shall return to these quotes in a few lines, and I shall freely assume that there is intrinsic e-value. Our question in this chapter is *where the intrinsic e-value comes from, and the conjecture we are going to develop and defend is that it derives from human curiosity or inquisitiveness.* The first quote from Goldman already suggested the connection, without making the crucial step of deriving value from curiosity. Similarly in the continuation of Alston's quote:

> [T]he attainment of knowledge and understanding are also of intrinsic value. "All men by nature desire to know," said Aristotle, and this dictum has been reaffirmed by many of his successors.

And he goes on in the same direction:

> Members of our species seem to have a built-in drive to get to the truth about things that pique their curiosity and to understand how and why things are as they are and happen as they do. So it is as close to truistic as we can get in philosophy to take truth as a good-making characteristic, and falsity as a bad-making characteristic, of beliefs and other outputs of cognition. (Alston 2005: 31)

The quotes point in the direction of the thesis that the desire for knowledge, or truth, or something similar is connected to the intrinsic value of these items, but they do not tell us what the connection is exactly like. Grimm himself is a bit more explicit:

> [A]ccording to this way of thinking, our curiosity about how things stand in the world is … importantly like the thirst we (characteristically, at least) feel when our body is dehydrated. When our body is dehydrated—when we experience thirst—*satisfying* our thirst is naturally thought to possess a kind of intrinsic value. (Grimm 2008: 727)

(Let me note that he does not clearly endorse the way of thinking he mentions.) I shall try to formulate the claim about the assignation of e-value by the desire to know and, in doing it, assign the central role to the desire, which I shall simply call "curiosity". Curiosity bestows e-value. It is central for the area of epistemology, and we shall be exploring one important aspect of this centrality.

Here is the brief preview. We have two issues to address, first, the metaphysical one: where does the e-value come from? Is it more subjective or more objective? Second, the epistemological issue: which items are really epistemically basic?

I want to start from the metaphysical issue, the crucial issue in epistemic axiology. So, in the next section, I shall be talking about grasping, leaving open its precise nature, shall go straight to the metaphysical issue of the source of value. I shall introduces the idea of response-dependence, and proposes that e-value is response-dependent. Graspings of "p" are e-valuable because they would be positively valued by relevant cognizer(s), on the basis of interest in whether *p*, or p-curiosity. Curiosity is the foundational epistemic virtue, since it bestows epistemic value.

Section "Targets of Curiosity: Bearers of Epistemic Value" discusses the issue of the target of idealized curiosity which is at the same time the fundamental bearer of epistemic value, namely a relatively minimalist kind of knowledge. Mere true belief cannot be rationally accepted in isolation from a supporting structure. This point is introduced by a discussion of coffee-machine thought experiment, and the e-value of reliability is affirmed and discussed. The result is then developed into a characterization of the structure(s) that are serious candidates for bearer of e-value.

Section "Conclusion: The Centrality of Curiosity Again" brings together the results from the two previous sections, and connects them to the conjecture that curiosity is also an organizing epistemic virtue. It thus ends with the double claim: curiosity organizes all other epistemic virtues, and it bestows e-value on our knowledge, and knowledge-like states.

Response-Dependence, Curiosity and Value

Once upon a time I wanted to know whether Neptune's moons are rich in water, and I learned from a reliable source that they are. Sue wanted to know which town is the capital of Rwanda, Inan tells us (2012: 138), and she learned that it is Kigali. I ended up having a piece of propositional true belief, hopefully even a piece of knowledge. Sue ended up with a piece of ostensible information ("Kigali") with which to replace her former inostensible description "the capital of Rwanda". We ended with positive "graspings", to use the term introduced in the previous section. I had a propositional grasping, G*p*, of the proposition *p*, Sue had an objectual one, G*o*, of the object *o*. We were both happy, since we reached something that has epistemic value (e-value, for short) for each of us. We shall discuss in the next section the precise nature of the bearers of e-value. Here, let me just mention that I agree with a long tradition that on the propositional side truth is paramount, and I assume that in the case of objectual acquaintance the correctness play the same role. But what about achievement? It is welcome but is not crucial.[3]

Now, how can we account for such intrinsic e-value? Someone may try to avoid the problem by denying the existence of intrinsic e-value: nothing is intrinsically valuable as the object of cognition, or as the state of grasping it. Some of my students were defending this intrinsic e-value nihilism. We agreed at the beginning of the chapter that nihilism is

---

[3] So many items are valued that do not involve significant achievement, as Duncan Pritchard has argued at length, in detail and to my mind convincingly (see his 2014 work). But note that he goes very far:

> When I say that truth is the fundamental epistemic good, I mean that from a purely epistemic point of view it is ultimately only truth that we should care about. Call this the truth thesis …

… Elsewhere, I have characterised this view as epistemic value T-monism, in that:

> (i) it is a view about epistemic value specifically (that's the "epistemic value" part);
> (ii) it says that there is just one finally epistemically valuable epistemic good (that's the "monism" part); and
> (iii) it says that this finally epistemically valuable epistemic good is truth (that's the "T" part). (2014: 114)

untenable: knowing facts about Neptune, or Rwanda, might be intrinsically valuable.

So, the tactics to be followed in this section will be to assume that some *Gp*s (and G*o*s) are e-valuable; where does the intrinsic epistemic value of G*p*s (and G*o*s) come from? Let me call both G*t* for short, "*t*" standing for "target". Now, what is epistemic valuing like? The usual feeling ("phenomenology") is clear: some G*t*s concerning some states of affairs (or objects) are intrinsically e-valuable, and people, if intelligent, well-informed-educated and sensitive are curious about these states of affairs. Here is Inan:

> If curiosity always involves interest and interest always involves values, then it follows that curiosity is always value-laden. I believe that such a position is correct. This would imply that strictly speaking there is no such thing as "sheer curiosity", if the term is taken to refer to a mental state in which one is merely curious about something that is not motivated by anything he values. (2012: 128)

And the link with the desire to know is easy to spot:

> If such an interest causes a desire to know, then it must be of the second order, in that the curious being not only has to be aware of what he or she does not know but must also desire to come to know the unknown. What exactly is involved in such a desire to come to know the unknown, how it is possible, and whether such a desire is to be taken as identical to that mental state of curiosity are issues to be explored now.
>
> It appears that the general tendency is to take curiosity as being an essential tool in achieving something that has intrinsic value, whether that is knowledge or understanding, in the propositional or objectual sense. (2012: 10)

This is what we have on the side of the desiring or interested cognizer. On the side of the object, the target to be grasped, grasping itself, or its external referent, we have a brute fact of being valuable. But why would a target of curiosity have *intrinsic epistemic* value? Take Neptune's water-rich moons and the capital of Kigali (or the fact that a determinate city is such a capital). Why would anything about them have *intrinsic epistemic* value? Consider first the extrinsic value: information about Neptune's water-rich moons is epistemically extrinsically valuable because it is useful *for us*. But the idea that there is a non-relational intrinsic value attached to them (or to Kigali), and that such a strange fact could be dictating epistemic axiology seems a bit extravagant. Just postulating that it has one leaves epistemic value unexplained.

We are thus facing Euthyphro's dilemma concerning the order of determination: does curiosity bestow value upon truths and graspings of truths or is it the other way around? Analogous questions arise about other kinds of value (moral, aesthetic) and the usual feeling ("phenomenology") and the kinds of options are the same. As we are all aware, there are roughly three groups of options altogether, differing in the order of determination.

First option: graspings (G*p*, G*o*) and their objects are not really valuable; "e-value" is mere projection. We might call it value nihilism, or strong anti-realism (projectivism). An example of it is offered by Stephen Stich in his 1990 book.[4]

Second option: graspings (G*p*, G*o*) and their objects are intrinsically e-valuable in themselves—strong realism. Here, the strong realist claims that intrinsic e-value determines human curiosity (at least in the right cognizers). A fine defense of such objectivism about e-value can be found in Michael Brady (2009) and I shall be addressing some of his arguments a few pages below.

Third option: graspings (G*p*, G*o*) and their objects are e-valuable because of our curiosity- dispositionalism or response-dependence view. This is the view to be defended here.

Note the analogy with color: strong anti-realism (projectivism) would claim that "nothing is really red". The strong realism that being red is completely objective feature red because it produces some relevant redness related perceptual state. Finally, response-dependence view has it that a surface is red because it tends to produce redness response in relevant observers under normal circumstance.

[4] See my extended criticism in Miščević (2000).

I shall set aside the strong projectivism that comes close to error theory, and e-value nihilism, and turn directly to response-dependentism (dispositionalism) and thereby to the claim that

Intrinsic curiosity is the e-value bestowing epistemic virtue. This is the strong (and, to many tastes problematic) claim that I want to start to defend here. Here is the general form for accounting for e-value:

> *p* (&G*p*) are e-valuable iff a person H, sufficiently cognitively normal (or, alternatively, idealized), and familiar with the domain of p, would be stably intrinsically curious about p.

Now we need a bit of refining. Let me start by introducing a distinction. When Sue learns about Kigali being the capital of Rwanda, she experiences the information (and her grasping of it) as being valuable. This is the subjective aspect, and I shall talk of e-value as experienced, or e-VALUE_{exp} distinguishing it from the objective e-value we want to account for (compare it to the experience of surface being colored, in contrast to the objective color). So, please note the terminology: Value as experienced = value_{exp}.

The experience represents the information about Kigali (and the grasping of it) as being valuable; the value as experienced is being felt as the property of the information (and grasping). I can feel how valuable this thing I have learned is for me, Sue might think. And the experience is transparent; it goes right to the target itself. The e-value is *transparently* present in the target:

> (1) e-value_{exp} is being experienced as being a property of a state of affairs. (*A Transparency Datum*)

The datum is both obvious and robust. It crucially distinguishes the experience of value from the experience of pain-causing devices. Locke's mana(a laxative inducing stomach pain), a device that produces pain in the thumb, say a thumbscrew, or an imaginary pain-producing surface, like those in Wittgenstein's thought experiment are, or would be, experienced in a quite different way. Victim's perceptual apparatus does not ascribe to those a phenomenal property corresponding to their pain-producing power. With value, as with color, things stand otherwise: they are experienced as belonging to the targets.

(1a) Intentionally experiencing e-value is an act of axiological intuition.

I use the term "intentionally" in the sense of being object-directed.

Remember our wondering at what non-human facts could make an item (like "Neptune's moons are rich in water", or the Kigali fact) intrinsically epistemically valuable. In a sense, the wondering points to a minimally naturalistic stance: there is nothing in the nature of physical reality that accounts for axiological properties. This gives us our next premise:

(2) The e-value$_{exp}$ is not an experiencer-independent property of the state of affairs. (*Naturalism*)

It has been objected by Stroud that accepting the scientific, "unmasking" premises, like our (2), leads the theoretician to believe there are no corresponding properties. His example involves color. He claims that in order to defend such view, the theoretician must be able to "identify perceptions as perceptions of this or that colour without himself ascribing any colour to any physical object", and this "cannot be done" (2002: 245; the argument is deployed at length in his 2000 work, Chapter 7). However, this objection underestimates the possibilities of bootstrapping: the unmasking theoretician starts in his own case with the full panoply of commonsense beliefs, and then proceeds by weakening them, as his theorizing progresses, going from "this is red" to "this looks red to me", where the content of "red" accordingly changes. To apply it to our case, the response-dependentist theoretician starts in his own case with the full panoply of commonsense beliefs, and then proceeds by weakening them, as his theorizing progresses, going from "this is valuable" to "feels valuable to me", where the content of "valuable" accordingly changes. And he does not have to end as a value nihilist, as we shall see in a moment.

(3) The e-value$_{exp}$ is not a property of subjective state. (*From Transparency*)

It is the value projected onto the target (moons, Kigali and facts about them). Unfortunately, both claims, (2) and (3), attribute a certain error to Sue. But, no one is perfect. And our everyday experiences and folk conceptualizations offer no guarantee of being error-free. Sue's error might be like the folk error of taking "up" and "down" as absolute properties of space. It is not dramatic, but it is an error nevertheless. *Charity* in interpretation dictates that we don't see folk as referring to nothing whatsoever when referring to directions conceptualized in the absolutist, folk way. Rather, they are best interpreted as managing to refer to the property that is the closest cousin to the intended one. The point is not just minimizing the error, but also rationalizing it, making it intelligible. Charity and *inference to the best explanation* go hand in hand. The traditional dispositionalist or response-dependentist thesis honors both. It captures the fact that *the closest actual referent for color concepts and expressions* is the disposition of surfaces to cause the target intentional states. And that *the closest actual referent for value concepts and expressions* is the disposition of targets to cause the right intentional states. And it does this stressing the right order of determination: what makes a surface red is its state-causing power, and not the other way around, what makes the info e-valuable is causing the satisfaction of curiosity, and not the other way around. Therefore (*by principles of charity and by inference to the best explanation*):

> (4) Conclusion: Being e-valuable in objective sense is being such as to cause the response of experiencing e-value$_{exp}$ in *normal/ideal-? observers* under *normal circumstances.* (*Response-intentionalism*)

Let us now start unpacking the Conclusion. As for observers, we have left open two options, the first referring to normal observers, the second to ideal ones. Start with the first, the egalitarian one:

> *p* (&G*p*) are e-valuable iff a person H, endowed with at least normal cognitive capacities and at least some general knowledge, and familiarity with the domain of p, would be stably intrinsically curious about *p* (either whether *p* is true, or about truths in connection with *p*, or both).

Assume that the cognizer is aware of her cognitive capacities (a small idealization). But now, why do we say "in connection with *p*"? To deal with the "curious facts" problem raised by M. Brady (2009: 278–279). Suppose we think about the following piece of information:

> It is forbidden for aircraft to fly over the Taj Mahal.

Brady suggests that we are happy to know such facts without having any antecedent curiosity about them. I suggest that there is a consequent curiosity: we appreciate grasping them because we find them curious, raising further questions, like why anyone would forbid flight over Taj Mahal and the like.

But this is just the beginning of a dialogue with Brady, who has come up with a collection of objections to the response-dependentist account in his paper on curiosity and the value of truth in the *Epistemic Value* volume. Here is his remark about the egalitarian version of the account (he doesn't call it "egalitarian" himself):

> But there seems to be a strong reason to be sceptical about this line on epistemic value. For it is a general truth in value theory that, although the fact that I do desire or care about something might incline us to think that that thing is worth desiring or caring about, it does not guarantee that it is. There is always the possibility that I desire or care about something that I ought not to desire or care about, that is, something that is not worthy of my concern. In other words, there is always the possibility that one of my ends or goals is not a proper end or goal. If so, we might think that the fact that I desire the truth on a particular subject for its own sake does not guarantee that the truth on that subject is worth desiring, or is valuable as an end. (2009: 269)

We obviously have to idealize; the question is how much. Here is the general form:

> G*p* is e-valuable iff a person H, endowed with (decent to high) cognitive capacities and general knowledge, and familiarity with the domain of *p*, would be stably intrinsically curious about p (either whether *p* is true, or about truths in connection with *p*, or both).

Michael Brady in his paper delineates such a position without endorsing it; in fact he proceeds to criticize it and ends by rejecting it. Here is the proposal:

> How can we move from the claim that we are naturally curious to discover the answers to particular questions, to the claim that answers to those questions are valuable in themselves? This problem is pressing, given that there might be something amiss with our curiosity or concern, and which therefore casts doubt upon the value of the truths which constitute the object of that curiosity or concern. A simple solution is to idealize the relevant concern for truth. Thus, we might claim that the truth on a certain issue is valuable, not if someone does care about or desire the truth on that issue, but only if the person would care about the truth under certain idealized conditions: if, for instance, the person would desire the truth on that issue were she fully rational(…). a process of rational idealization will bring to light whether the subject's interest is instrumental or intellectual, will ensure that inquiries are not based upon false beliefs, and will rule out curiosity that results from irrational compulsions. We might therefore maintain that it is the satisfaction of natural and rational, idealized curiosity which has final value. (2009: 271)

Obviously, the proposal needs a lot of work to arrive at the right level of idealization. Too little is unsatisfactory, given human limitations; unfortunately, some people are intrinsically curious about worthless matters. Too much is equally bad: only high level problems will be intrinsically interesting to such epistemically ideal person. In addition, we have the issues of depth and width: short of omniscience, what is the right proportion of going into detail and depth, and wanting to encompass as many areas as possible. So, the general question is with us, concerning both subjects and circumstances: how much idealization and of what kind? I still believe that intrinsic curiosity is the e-value bestowing epistemic virtue. Instead of trying to solve all the difficulties at once, I shall limit myself to a handful of problems, some of them raised by Brady and his original and challengeng counterexamples. Some of them we have already addressed in the preceding two chapters, so I shall be brief about them.

First problem: the superficiality of novelty. In his "Interest and Epistemic Goodness" (2011) Brady starts from psychology: "There is wide agreement—among psychologists, at least—on the appraisal variables that generate interes." He writes, "One of the central appraisals is of novelty: 'whether or not an event is new, sudden, or unfamiliar. For interest, this novelty check includes whether people judge something as new, ambiguous, complex, obscure, uncertain, mysterious, contradictory, unexpected, or otherwise not understood'" (Silvia 2006: 57). He also mentioned that interest and importance diverge. In the handout he points out that "we tend to find old, expected, familiar things comfortable or enjoyable, but are interested in things which are unexpected, unfamiliar, mysterious, baffling" (p. 2). So, the curious person starts by noting that something is ambiguous (complex, obscure, mysterious, contradictory), and asks oneself how one should one understand it. He finds such interest superficial and unstable.

Answer: to me it seems that if curiosity is directed to the "new, ambiguous, complex, obscure, uncertain, mysterious, contradictory, unexpected, or otherwise not understood". Then its central goal is achieving understanding, rather than arriving at isolated items of knowledge, and I think it is epistemically quite a good thing. The interest in complexity leads to the desire to understand, the crucial epistemic desire. Novelty is in the vicinity; it involves not-yet-understood matters. Finally, a virtuous researcher is able to control herself, to balance novelty with relevance and depth, and so on. So much for the first line of defense. But one may also add that the interest in the novel and the complex is, globally seen, extremely epistemically useful. The novelty liberates us from cognitive inertia; just think of depressed people who have lost their natural curiosity.

Second problem: M. Brady's symmetrical problem for curiosity as source of value:

> /t/here are *epistemic* windfalls, truths whose value depends upon the fact that they were unsought, and so depends upon the fact that they were not the results of inquiry. (2009: 280)[5]

[5] Here is a longer quote:

> [W]e might think that there are epistemic windfalls, truths whose value depends upon the fact that they were unsought, and so depends upon the fact that they were not the results of inquiry. For example, if unsolicited affection constitutes a positive value in our lives, we might think that unsolicited knowledge of affection does as well. Thus, I might learn that "she loves me" because of her

He claims that, for instance, unsolicited knowledge of affection constitutes a positive value in our lives.

Answer: suppose I care for love of three persons, Jane, Julia and Peter, but I don't care at all whether Kate loves me, and I don't give a damn for info about it. Why would "unsolicited knowledge" of her affection constitute a positive value in my life? So, i assume that these counterexamples to response-dependentist account do not really threaten it. On the other hand, if I cared about Kate's feelings, I would have normally asked myself whether she has affection for me, and thus I would have been (perhaps very passively and lazily) curious about the matter).

The third problem: the fact-value gap. Here is a remark against response-dependentism made by Stratton-Lake in his introduction to Ross' classic The *Right and the Good*:

> /o/ur knowledge that certain things are intrinsically good does not seem to be derived from other evaluative knowledge, and given the autonomy of ethics, this knowledge cannot be derived from non-evaluative premisses, such as our knowledge that we desire or approve of that thing. (xliii)

A simple answer: let us accept for the sake of argument that moral value is completely autonomous. We have no reason to accept analogy with e-value; it is simply not so separate from its factual supervenience basis as moral value is.

The fourth problem: omniscience. Inan, Carter, and my student M. Bakalova warned me that a person, who knows everything and is thus epistemically close to perfection, would not be curious, and would thus paradoxically lack the alleged main motivating epistemic virtue,

One answer is that many human virtues are tailor-made for human agents in less-then-perfect but better-than hellish human circumstances. Curiosity is one such virtue, typical for finite and relatively ignorant beings, in need of constant updating of information in order to function

---

unsolicited declaration of love. Here my true belief has value that it would lack if it resulted from inquiry on my part. There seem to be a great number of surprising but welcome truths that fall into this category. So the efforts of inquiry are sometimes incompatible with the intrinsic value of true beliefs. (Ibid.)

successfully (analogy: an all-powerfu, even omnpotent being does not need courage).

But I would add more: I just to stipulate a slightly wider meaning of "inquisitiveness" that also includes cherishing the truth once found. It seems to me a natural extension of the narrower meaning: a person with bad memory but eager to get to know, who subsequently doesn't care a bit for the knowledge acquired and is completely unworried about having forgotten everything she learned, is not consistently inquisitive. So, the hypothetical omniscient person who keeps her virtue by cherishing what she knows is "curious" in this wider sense.

The fifth problem is the issue of bad curiosity: some cases of curiosity are really bad. How can curiosity then bestow any positive value?

Answer: most bad curiosity is the one that is extrinsically motivated (envy, bad goals, etc.); but what if I am intrinsically motivated, but my curiosity is still unacceptable (say, curious about the private life of my student, whom I just find an interesting person, without having further goals but grasping truths about him)? In these cases, the moral disvalue (in the example, the derogation of privacy) counterbalances the intrinsic e-value, and wins (there are two further sub-options: either the e-value is annihilated, or it stays there but is simply defeated by the negative extrinsic, moral disvalue).

Sixth problem: sometimes intense curiosity can block the insight. Scientists tell us that they got their best ideas when they stopped being obsessed with the issue they were working on; suddenly the insight would come, often in unexpected circumstances.

Answer: psychologists agree with scientists-discoverers, but they tell us that the best explanation is to postulate the existence of a sub-personal inquisitive drive; see the book *The Eureka Factor: Aha Moments, Creative Insight, and the Brain* by John Kounios, Mark Beeman (2015).

There are further issues to be addressed: kinds of curiosity reflected on the features of e-value, the nature and the origin of e-disvalue, and many more. But we have to conclude. Let me reiterate the main idea of the section and of the chapter: intrinsic curiosity is the e-value bestowing epistemic virtue. Probably *most things that concern us in our normal human lives are response-dependent*, goodness versus wickedness, beauty versus ugliness, attractiveness versus repulsiveness, being humanly meaningful versus being meaningless and empty. In contrast, *most things that are metaphysically important are not response-dependent.* To put it in a form of a slogan, *response-dependence belongs to the manifest picture we care about humanly, independence belongs to the deep reality we care about scientifically.* Philosophy is the happy branch in which we can discuss both.

Let me now turn to the empty slot I left in the story. What are the targets of curiosity and the bearers of epistemic value? Although I think that the proposed account would work for a very wide range of candidates, an opponent might see the lack of discussion of the topic as a fatal lacuna in the account. So, the question should be addressed. It will take a lot of space, in comparison with the main topic, but still I apologize for too brief a treatment of an intricate and important topic.

Targets of Curiosity: Bearers of Epistemic Value

Why Is Reliability Valuable? The Swamping Problem and the Coffee-Machine Thought Experiment

We have been freely talking about "grasping" as candidate bearer of epistemic value. But what kinds of doxastic-epistemic states are eligible candidates? Let us stay with propositional curiosity and corresponding states; we shall try to generalize our result(s) to their objectual counterparts later. Certainly, we have true belief, (internally) justified true belief, knowledge, understanding and perhaps even more, for example, wisdom. Does each item have a value? And what are the paramount qualities that support the value. Let me agree with a long tradition that truth is paramount. But what about achievement? It is welcome but is not crucial. So many items are valued that do not involve significant achievement, as Duncan Pritchard has argued at length, in detail and to my mind convincingly (for instance, in his 2014 work).[6] So it is good to have true

[6] But Pritchard goes very far:

> When I say that truth is the fundamental epistemic good, I mean that from a purely epistemic point of view it is ultimately only truth that we should care about. Call this the truth thesis …

belief, and to have internally justified true belief, and reliably acquired true belief, and knowledge. But also the components (justification, reliability) seem to be valuable. Knowledge seems to have a high status partly because of its stability and reliability. On the theoretical level it would be nice to have an account that could order the bearers of e-value, for instance, show that value of understanding is bigger than value of knowledge that is bigger than value of justified true belief that is bigger than value of mere true belief (what about reliably acquired true belief). But some comparisons might be difficult, and there might be no consensus about ordering.[7]

So, let me start by discussing the value of stable, reliable origin. It has been famously contested by Linda Zagzebsky, for instance, in her 2003 paper, where she offers a few remarks on coffee and coffee machines, that have been reconstructed as a provocative thought experiment. I shall use the summary offered by Duncan Pritchard in his (2011), since it makes clear the thought-experimental character of the argument:

> Imagine two great cups of coffee identical in every relevant respect—they look the same, taste the same, smell the same, are of the same quantity, and so on. Clearly, we value great cups of coffee. Moreover, given that we value great cups of coffee, it follows that we also value reliable coffeemaking machines—i.e. machines which regularly produce good coffee. Notice, however, that once we've got the great coffee, then we don't then care whether it was produced by a reliable coffee-making machine. That is, that

... Elsewhere, I have characterised this view as epistemic value T-monism, in that:

(i) it is a view about epistemic value specifically (that's the "epistemic value" part);
(ii) it says that there is just one finally epistemically valuable epistemic good (that's the "monism" part); and
(iii) it says that this finally epistemically valuable epistemic good is truth (that's the "T" part). (2014: 114)

[7] Compare John Gibbons' (2013) book on the norm of belief. He notes that the following are all fairly plausible claims about when we ought to believe things:

(T) You ought to believe p only if p is true.
(J) You ought to believe p if and only if you're justified in believing p.
(K) You ought to believe p only if you'd thereby know that p.

And that though they're all plausible, they can't all be true. But, he tries to do justice to all of them.

> the great coffee was produced by a reliable coffee-making machine doesn't contribute any additional value to it. In order to see this, note that if one were told that only one of the great identical cups of coffee before one had been produced by a reliable coffee-making machine, this would have no bearing at all on the issue of which cup one preferred; one would still be indifferent on this score. In short, whatever value is conferred on a cup of coffee through being produced by a reliable coffee-making machine, this value is "swamped" by the value conferred on that coffee in virtue of it being a great cup of coffee. What is the swamping problem. (2011: 246–247)

Pritchard calls it swamping argument and here is his formulation:

(1) The epistemic value conferred on a belief by that belief having an epistemic property is instrumental epistemic value relative to the further epistemic good of true belief.

The second claim is the general thesis about value:

(2) If the value of X is only instrumental value relative to a further good and that good is already present, then it can confer no additional value.
(3) Knowledge that p is sometimes more epistemically valuable than mere true belief that p. (Ibid.)

This brings the Swamping Problem onto the scene: if the value of a property possessed by an item is only instrumental value relative to a further good and that good is already present in that item, then this property can confer no additional value on that item. This holds for epistemic properties in relation to the good of truth. So, knowledge that *p can be no more valuable than mere true belief that p*. Pritchard accepts (1) and (2) and rejects (3). Knowledge has *no added epistemic value* in comparison to true belief. Justification is epistemically worthless! But this seems really counterintuitive and problematic. Is there a way out?[8]

[8] Here is a longer quotation from J. Kvanvig offering an analogous problem. He talks about the Meno problem, of whether and, if yes, why knowledge is more valuable than true belief:

> Assumption 1: The Meno problem can be solved if there is a property P that (i) distinguishes knowledge from true belief and (ii) is a valuable property for a belief to have.

Let us look at the coffee thought experiment again. Are we ever being offered the choice as described?

> Here are two beliefs, for example:
>
> (1) *Wuhan is in China.*
> (2) *Maribor is the Slovenian town closest to Graz.*
>
> Both are true, but (1) is from an unreliable source and (2) is from a reliable one. Which one do you prefer?

Did you ever receive such an offer? Does it make sense? Imagine: I am telling you that Maribor is the Slovenian town closest to Graz, and that you are hereby getting it from an unreliable source! If I am offering you the choice, and you can trust me that "*Maribor is the Slovenian town closest to Graz…*" is true, then you are getting your belief from a reliable source. If you cannot, the offer cannot be formulated. In short, *there is no viable equivalent of the tasting of coffee, no neutral checking*: if the checking is worthy of its name, it yields more than mere true belief: either justified true belief, or knowledge. If it does not, it does not test for the truth of the belief. So, the coffee thought experiment is ok for coffees. But it lets

Assumption 1, however, is false. To see that it is false, consider some simple analogies. If we have a piece of art that is beautiful, its aesthetic value is not enhanced by having as well the property of being likely to be beautiful. For being likely to be beautiful is a valuable property because of its relationship to being beautiful itself. Once beauty is assumed to be present, the property of being likely to be beautiful ceases to contribute any more value to the item in question. Likelihood of beauty has a value parasitic on beauty itself and hence has a value that is swamped by the presence of the latter. Take anything that you care about: happiness, money, drugs, sports cars, and so on. Then consider two lists about such things, the first list telling you where to obtain such things and the second list telling you where you are likely to obtain such things. Now compose a third list, which is the intersection of the first two lists. It tells you of ways and places that both are likely to get you what you want and actually will get you what you want. But there would be no reason to prefer the third list to the first list, given what you care about.

These analogies show that when the value of one property is parasitic on the value of another property in the way that the likelihood of X is parasitic on X itself, the value of the first is swamped by the presence of the second. So even if likelihood of truth is a valuable property for a belief to have, adding that property to a belief already assumed to be true adds no value to the resulting composite that is not already present in true belief itself. So Assumption 1 is false; one cannot solve the Meno problem simply by finding a valuable property that distinguishes true belief from knowledge. (Kvanvig 2003: 45) (Thanks to J. Adam Carter for pointing the passage out to me.)

us down at the stage of generalizing (all kinds of goods) and of analogizing beliefs to cups of coffee. Morals: the coffee model is not applicable to beliefs

I have been telling the story in terms of propositional knowledge, but it can be retold in terms of objectual curiosity and knowledge, dear to Inan. Let us repeat the game. I just told you:

> Here are two sentences
>
> (1) *Wuhan is in China.*
> (2) *Maribor is the Slovenian town closest to Graz.*
>
> Both are true, but (1) is from an unreliable source and (2) is from a reliable one. Which one do you prefer?

Consider now the critical definite descriptions "the country in which Wuhan is located" and "the Slovenian town closest to Graz" You have started with two inostensible concepts, the first corresponding to "the country in which Wuhan is located", and the second corresponding to "the Slovenian town closest to Graz". In the game I am also offering you their ostensible equivalents, "China" and "Maribor", but I am doing it in a thoroughly unacceptable way, by saying that the first offer is reliable and the second is not. But it makes no sense to make an offer and then claim it is unreliable. It is not like offering two coffee cups that taste the same. The analogy with coffee fails for the ostensible/inostensible contrast as well as for the more traditional epistemological concepts.[9]

---

[9] Here is the third consideration, due to Frederick F. Schmitt and Reza Lahroodi:

BELIEF-MACHINE VARIANT The person who is curious whether Goldbach's Conjecture is true would not be fully satisfied by a mere true belief as to whether it is true. If offered a choice between a device that would, upon pressing a button, implant a true belief as to whether the Conjecture is true and a device that would implant knowledge, the subject would prefer the latter device and would do so to satisfy curiosity. Indeed, the requirement of knowledge is not merely for a justified true belief. ("The Epistemic Value of Curiosity" (2008), *Educational Theory*, Volume 58, Number 2)

So far, so good. But it is too little to say that the subject would just prefer the knowledge machine.

Imagine waking up with the mere belief: Goldbach's Conjecture is true. No reasons, no awareness of the source! Like the Truetemp. It would be quite irrational to accept the belief-machine offer.

So, merely true belief taken in isolation cannot really be rationally accepted. Belief is unlike coffee in crucial respects. Most importantly, its value cannot be tested without the test importing new, crucial information that turns true belief into something more powerful (justified true belief or knowledge). Therefore, a *de facto* true belief cannot be rationally accepted in isolation from these crucial additions.

The coffee model seems to make sense from the 3rd person perspective, but not from the first-person perspective of the cognizer. And it is the cognizer that is being asked about her preferences, not the external judge. But what kind of stability is involved here? Pritchard claims it is merely practical,[10] but in our context, there is no mention of practical use. It is a matter of pure credibility. so, it should be *informational or epistemic*. Belief is unlike coffee in crucial respects. Most importantly, its value cannot be tested without the test importing new, crucial information, that turns true belief into something more powerful (justified true belief or knowledge). Therefore, a *de facto* true belief cannot be rationally accepted in isolation from these crucial additions. Let me generalize. Here is a general dilemma for the coffee model:

> If you were informed from an epistemically authoritative source that "p" is a true belief, then you would have reliable information that p.
>
> If you were not, than you would have no reason to accept that p.

---

[10] Pritchard writes:

> /t/here is little to be gained by responding to the swamping problem by arguing that the epistemic standing in question generates a practical value that mere true belief lacks. For example, suppose one responded to the swamping problem by arguing that knowledge is more valuable than mere true belief because knowledge entails justification and justification is practically valuable. Justified true belief, we might say—in a broadly Socratic fashion [...]—a "stability" that mere true belief lacks, and this means that it is more practically useful to us in attaining our goals. The problem with this response, however, is that it doesn't appear to engage with the swamping problem at all. After all, the difficulty that the swamping problem poses concerns how to make sense of the idea that knowledge is more valuable than mere true belief because it involves an epistemic standing which better serves our specifically epistemic goals—in particular, the epistemic goal of true belief. Thus, the kind of value that is at issue is specifically an epistemic value. Accordingly, even if it is true that knowledge has more all-things-considered value because it entails an epistemic standing which adds practical value to true belief, the problem would still remain that, on the face of it, knowledge is not epistemically more valuable than mere true belief. (2011: 246–247)

There is no middle ground here. The opponent, for instance, Pritchard, might try to argue that our point is simply a matter of pragmatics. Indeed, offering a piece of information and claiming at the same time that the source, namely the speaker herself is unreliable is a pragmatic contradiction, or paradox. But this pragmatic incompatibility tells us nothing about the actual distribution of e-value, he might claim. We need an example of a situation where pragmatic considerations are blocked, but the importance of justification and reliability remain. Here is a possible example:

> The elections
>
> You are very curious about the presidential elections in my country, which involve two candidates, Kolinda and Josip. It is the election day, 5 p.m. The results are not yet known, they will be known at midnight, but you are not aware of it, and you trust me.
>
> I try a practical joke. I toss a coin (at 5 p.m.) And the coin says "Kolinda". I call you and tell you "Kolinda is the winner". You accept, form the belief and you thank me warmly for the info.
>
> At midnight, it becomes public that Kolinda indeed won. I call you, and tell you that it was a joke, and I had no clue when I called you. "But at least, my info was true", I add. How would you react?

One rational reaction: "Well, don't do it again, Nenad!" Others would be along the same line, criticizing me for my stupid joke. Suppose I answer:

> Yes, but your belief was true, you should appreciate it a lot by your own lights!

This is even worse. It looks like the worth of merely true belief is rather minimal. And it looks that by making you accept the true belief that Kolinda is the winner, I did you a disfavor.

Moral one: true belief is valuable, but implanted alone it has a minimal value!

The impression can be strengthened: we generally don't regard a stable arrangements as a series of one-shot deals: a good relationship is not a two thousand and one night stand, a stable home is not a series of many 24-hour-lasting improvised shelters. But with knowledge, it is even more dramatic. The one-shot offer itself does not make sense. Acceptability and reliability go together in a package deal.[11] I shall call the morals of the election story "package deal argument".[12]

I have been telling the story in terms of propositional knowledge, but it can be retold in terms of objectual curiosity and knowledge, dear to Inan. So, in the story retold, you are interested in who the new president of Croatia is. You have an inostensible description of him/her, namely "the new president". What you want is a more ostensible information, let say the name (with all the problems that go with it, listed and brilliantly analyzed by Inan in his 2012 work (142 ff), in connection with the name "Kigali"). Now, with the practical joke I actually gave you the right information, its Kolinda. Still, you are not satisfied, after you hear about my actual ignorance at the time of giving the info. What is needed is the package deal: *ostensible information with some guarantee of reliability.* I cannot defend the fully isolated true belief (except going the Martin Luther WAY: here i stand and believe, Ich kann nicht anders!).

So, here is my proposal: combine the package deal argument with the failure of the coffee thought experiments. The resulting picture will be the following:

> Truth is the primary goal, but mere true belief is not the fundamental bearer of e-value. Rather, the bearer is a relatively undemanding, minimalist kind of knowledge. Curiosity follows the same pattern: a rational cognizer wants truth plus supporting structure.

Mere true belief is only minimally valuable for the curious cognizer. I told you the name of winner, you got the true belief, by pure luck. Truetemp got one by insertion into his brain. How valuable, epistemically speaking, is it for you and for Truetemp respectively? Not much; very little has been

[11] The type of combination is widespread, way beyond the mere intrinsic e-value of truth. Imagine you would value a lot having a nice drink. And you are offered a glass, you drink it and enjoy it. Next day you are told that it could have been poison. You would not thank the person for the nice drink, although the drink is what you basically value.

[12] Compare Carter et al. (2013).

given to you and to him. (You have right to be offended at my playing games with you, Truetemp at tampering with his brain, for very little in terms of *epistemic* gain!) So this is the typical epistemic value of true belief without supporting structure. It is not impressive. Plato already knew it: such true beliefs are like Socrates' *daidaleia*, moving statues-robots, utterly defenseless and ready to run away (*Meno* 97a–98b). Mere true belief cannot be rationally sustained in the face of a slightest bit of contrary evidence. For example, I believe (truly) it is not raining, but I have no supporting structure for my belief. A mere drop of water on my window, say from my neighbor's hose, makes me change my mind, and the true belief is gone. But also, my change of mind is in a sense less than rational. In contrast, if I have a supporting structure (I can see no clouds in the sky) the rational defensibility is there. Now, is rational defensibility merely practical and instrumental? Why would it be? Why is this not epistemic?

If you already have intuition that the additional element of stability and defensibility does add epistemic value, you can use a Modus Tollens: the additional element cannot add epistemic value unless it is itself epistemically valuable. It does add epistemic value. Therefore, the additional element is epistemically distinctly valuable:

Mere true belief (as well as mere correct ostensive information) cannot be rationally accepted in isolation from a supporting structure.

However, any efficient supporting structure introduces further epistemic goods (justification, reliability, anti-luck guarantees), thus upgrading the original true belief.

Mere true belief (as well as mere correct ostensive information) can be neither defended, nor rationally sustained through time, due to the isolation (see Carter et al. 2013).

Mere true belief cannot be rationally sustained in the face of a slightest bit of contrary evidence (the Meno insight). Therefore:

Mere true belief (as well as mere correct ostensive information) is not rationally stable.

Mere true belief (as well as mere correct ostensive information) is only minimally valuable for the curious cognizer.

Epistemic goods come in package deals.

Rational stability is an epistemic, not merely a practical property (or status).

Let us leave open how massive the supporting structure should be. For our purposes a molecular, not holistic structure is enough. The Truetemp analogy suggests that the structure should contain an indication of origin, some indication of circumstances (perceptual, testimonial, memory-based belief). All this might help to account for the value problem. Let me just not the direction of solution, leaving the details for another occasion. First, showing that knowledge is more valuable than mere true belief, which we did. Second, showing that knowledge is more valuable than that which falls short of knowledge. Justified true belief without some indication of reliability is not a satisfactory package deal.[13] Finally showing that knowledge is more valuable than that which falls short of knowledge not merely as a matter of degree but of kind. A very modest proposal: the special status comes from the fact that minimal knowledge is the first, or the basic kind of grasping the truth that has all the requisite qualities.[14]

Let me put my cards on the table in matters of the source of e-value of various candidates: the intrinsic e-value of true belief derives from the desire for truth, the intrinsic e-value of justified true belief derives from need of reflective certainty, and ability to defend one's belief and transmit it if needed. The need of reflective certainty, I submit, is epistemic as is the need of ability to defend one's belief and transmit it if needed (social epistemic). E-value of knowledge we discussed, it derives from all the preceding elements plus defensibility and stability (achievement is optional). It is probably the first satisfactory package that gives one, from the first-person perspective the epistemically stable supporting structure. Understanding is the next; its intrinsic e-value derives from its richness

[13] I leave for some other occasion the discussion of the view, due to Kvanvig, according to which knowledge is not the first inquiry stopper, whereas the gettierized justified true belief already is.

[14] All this should be argued for on the bases of various proposed accounts of knowledge. I've been stressing stability. But similar considerations hold for other proposals. Consider D. Pritchard's recipe for knowledge: virtue + anti-luck. The virtuous origin (like the old-style justification) secures the rationality of forming and keeping alive the belief. The anti-luck component caters for the stability.

and cognitive relevance and role in manipulating causes. All this would demand a lot of arguing; I have to stop here.

So much for beliefs and curiosity in general. But what about truly foundational beliefs (if there are any); where does package deal come from in their case? For instance, Wittgenstein's hinges? A possible answer is that they are presumably widely shared in the epistemic community ("shared" in several relevant senses), and their special status accompanies them as part of their package.

Next, what about the sub-personal level? I assume the story is roughly similar. Our cognitive modules trace the origin and credentials of various inputs. A normally functional cognitive apparatus is able to distinguish sub-personally imagined from sub-personally perceived contents. Let me borrow a pair of terms from E. Sosa (2015: 67 ff). He talks about biological—functional versus intentional, noting that on the biological level the proper function of human belief-system is to represent reality-as-it-is: the representation should be as accurate as possible given the costs. On the intentional level the proper function links beliefs to the truth-goal. My own preference is to think that intentional is continuous with biological (Dretske, Millikan), but I will not be dogmatic here. A broad parallelism will be enough. On the sub-personal level our cognitive mechanisms search epistemic stability-defensibility as much as on the personal level. So, there is no principled problem.

Conclusion: The Centrality of Curiosity Again

In this chapter I have tried to do two things concerning the value of truth and knowledge, and their relation to curiosity. First, and most importantly, to address the Euthyphro's dilemma concerning the order of determination: does curiosity bestow value upon truths and graspings of truths or is it the other way around? Second, to offer a sketch concerning the bearer(s) of epistemic value, and to adjudicate between purely truth-centered proposals, and wider options, including properties like reliability, stability and justifiedness.

Let me focus upon the first task. The chapter argues for a response-dependentist account of intrinsic epistemic value of true grasping (belief, knowledge): intrinsic curiosity is the value-bestowing epistemic virtue. In short: the value is normally experienced as being a property of a state of affairs to be grasped. However, value naturalism suggests that it is not an experiencer-independent property of the state of affairs. Hopefully, the value is not merely a fiction. Therefore, by principles of charity and by inference to the best explanation, being epistemically valuable in objective sense is being such as to cause the response of intentionally experiencing epistemic e-value in suitable under suitable circumstances. Our graspings of propositions and objectual characterizations are epistemically valuable if a person, endowed with at least normal cognitive capacities and at least some general knowledge and familiarity with the domain of *p* (or, alternatively, the person's somewhat idealized counterpart), would be stably intrinsically curious about *p* (either whether *p* is true, or about truths in connection with *p*, or both). Similar conditions hold for objectual curiosity. We have tried to address a number of objections to this view, and we hope to have offered at least beginnings of a right response. We concluded that curiosity is the foundational epistemic virtue that bestows epistemic value to its targets. Now, I would like to try connecting the claims to my previous work on curiosity.

I have tried in the preceding chapters to defend the following claims: first, that intrinsic curiosity is an epistemic virtue. Second, that it organizes and mobilizes other virtues, both abilities related and morality-related ones. Obviously, curiosity is *not* an ability, it is a *motivating* truth-seeking virtue, a choice-related feature of the mind, of the sort similar to generosity and courage. These virtues are normally praised by thinkers like Zagzebski who stress the motivating role of virtues. Curiosity also helps integrating other moral-like virtues in the picture and accounting for them. They are of two kinds. Either, they are directly aiding curiosity, like open-mindedness does, perhaps preventing the cognizer's mind to get clogged by worthless old stuff. Or, they have to do with other values (e.g. originality with the value of being new in an interesting way) and other kinds of virtue, above all moral virtues (e.g. generosity). One should see them as hybrids, partly moral, partly purely epistemic. This fits the intuition that they have high moral relevance, as well as the assumption that they favor reaching purely epistemic goals. This preserves both *primacy of truth-goal and the traditional and ordinary understanding of virtue as a motivating feature*. The result would be an integrated virtue-based view. What about cognitive capacities or capacity-virtues, like, for instance, well-functioning and well-integrated perception and rational intuition, the kind of virtues mentioned by Sosa and Greco inside their very definitions of knowledge? Are they really virtues? Yes, they are, in their own modest way and the truth-camp philosopher should not worry. However, they are not motivating virtues. They are *executive* virtues. They lead the agent to the epistemic goal set primarily by her inquisitiveness, pure or practical.

The proposal perhaps merits to be characterized as an *integrated virtue-based view*, since it is strongly aretaic, integrates motivating and executive virtues, and aims at seamlessly integrating the typical pursuits of virtue epistemology with the traditional business of epistemology. The character-virtue tradition and the truth-centered one can be married in a quiet and civilized fashion, without forcing any shotgun wedding between them. Combined with the present claim about the response-dependent nature of epistemic value, the proposal becomes even stronger: curiosity is *the central and the foundational* epistemic virtue. It is foundational since it bestows epistemic value, and central since it organizes other epistemic virtues.

The second issue is the one of the fundamental bearer of epistemic value. Truth is central for human cognitive-epistemic effort. I have argued, briefly and all too briefly, that truth is the primary goal, but that mere true belief is not the fundamental bearer. Rather, the bearer is a relatively minimalist kind of knowledge. Mere true belief cannot be rationally accepted in isolation from a supporting structure. However, any efficient supporting structure introduces further epistemic goods (justification, reliability, anti-luck guarantees), thus upgrading the original true belief. Keep in mind how little epistemic value commands the mere true belief (or mere correct ostensible presentation) without the supporting structure. And how much more, intuitively seen, is provided by justification and knowledge (and their objectual correlates). And note that the surplus comes from them alone, not from the minimal e-value of true belief.

Mere true belief can be neither defended, nor rationally sustained through time, due to the isolation. Mere true belief cannot be rationally sustained in the face of a slightest bit of contrary evidence (the Meno insight). Therefore, mere true belief is not rationally stable. Minimal knowledge is, and this accounts for value problem in its various guises.

On the side of objectual curiosity, we have similar candidates for the bearer of epistemic value besides mere correct ostensible presentation (concept), namely justified correct ostensible presentation and justified correct ostensible presentation with reliable underpinning, not to speak of understanding as a further candidate. As in the case of propositional belief, here epistemic goods come in package deals.

Let me reiterate: curiosity is *the central and the foundational* epistemic virtue. I hope this idea gives a general epistemological framework that would be very friendly to research on the inner nature and proper definition of curiosity.

References

Alston, William. 2005. *Beyond Justification: Dimensions of Epistemic Evaluation*. Cornell University Press.

Brady, Michael S. 2009. Curiosity and the Value of Truth. In *Epistemic Value*, ed. A. Millar Haddock and D. Pritchard, 265–284. Oxford University Press.

Carter, J. Adam, Benjamin Jarvis, and Katherine Rubin. 2013. Knowledge: Value on the Cheap. *Australasian Journal of Philosophy* 91: 249–263.

Gibbons, John. 2013. *The Norm of Belief*. Oxford University Press.

Goldman, Alvin I. 1999. *Knowledge in a Social World*. Clarendon, Oxford.

Grimm, Stephen R. 2008. Epistemic Goals and Epistemic Values. *Philosophy and Phenomenological Research* 77 (3): 725–744.

Hazlett, A. 2013. *A Luxury of the Understanding: On the Value of True Belief*. Oxford University Press.

Hurka, Thomas. 2001. *Virtue, Vice, and Value*. Oxford University Press.

Inan, Ilhan. 2012. *The Philosophy of Curiosity*. Routledge.

———. 2014. Curiosity, Belief and Acquaintance. In *Virtue Epistemology Naturalized: Bridges between Virtue Epistemology and Philosophy of Science*, ed. A. Fairweather, 143–158. Springer.

Kounios, John, and Mark Beeman. 2015. *The Eureka Factor: Aha Moments, Creative Insight, and the Brain*. Random House.

Kvanvig, Jonathan L. 2003. *The Value of Knowledge and the Pursuit of Understanding*. Cambridge University Press.
Lynch, Michael. 2004. *True to Life: Why Truth Matters*. MIT Press.
Miščević, Nenad. 2000. *Rationality and Cognition: Against Relativism-Pragmatism*. University of Toronto Press.
Pritchard, Duncan. 2011. What is the Swamping Problem? In *Reasons for Belief*, ed. Andrew Reisner and Asbjørn Steglich-Petersen, 244–259. Cambridge University Press.
———. 2014. Truth as the Fundamental Epistemic Good. In *The Ethics of Belief: Individual and Social*, ed. J. Matheson and R. Vitz. Oxford University Press.
Silvia, Paul J. 2006. *Exploring the Psychology of Interest*. Oxford University Press.
Sosa, Ernest. 2015. *Judgment and Agency*. Oxford University Press.
Stroud, 2002. Understanding human knowledge, Oxford University Press
Zagzebski, Linda. 2003. Intellectual Motivation and the Good of Truth. In *Intellectual Virtue: Perspectives from Ethics and Epistemology*, ed. Michael DePaul and Linda Zagzebski, 135–154. Oxford University Press.

Part III

Applications and Widenings

8

Cognitive Psychology of Curiosity

Introduction

Now, what about the science to be used for the task of accounting for human curiosity?[1] We shall look at cognitive psychology and concentrate on the tradition of appraisal theories, which propose that it is the appraisal of situations that causes or arouses curiosity in people: if I appraise the appearance in the corridor of philosophy department of a gentlemen that I don't know as something new to me, it will arouse my curiosity: who is this new person? The psychologist who originated the tradition has been Daniel Berlyne (see his 1954, 1960 and 1966 works). The approach, which many find the most promising in the present cognitive scene, has been further developed by Paul J. Silvia, whose main book appeared in 2006. Silvia's publications stretch from 2006 to the present day,[2] and he publishes in collaboration with very well-known colleagues working on curiosity, like Todd B. Kashdan (see, for instance, Kashdan and Silvia (2009). Silvia's work has already been used by philosophers; for us the

---

[1] Thanks go to my student Fatma Betul Aydin, with whom I have discussed Silvia quite a lot.

[2] See, for example, his 2018 "Everyday Creative Activity as a Path to Flourishing", T. S. Conner, C. G. DeYoung, P. J. Silvia, *The Journal of Positive Psychology* 13(2), 181–189.

N. Miščević, *Curiosity as an Epistemic Virtue*, Palgrave Innovations in Philosophy,
https://doi.org/10.1007/978-3-030-57103-0_8

most interesting pieces are Brady's papers and handout (e.g. "Curiosity & Epistemic Goodness" (handout); see the section on appraisal below).

Let me mention a possible reservation. Is Silvia's work really relevant? Yes, it is, for instance, quoted in Inan et al.'s book on moral psychology of curiosity, both by psychologists (Megan Haggard, 148 ff) and by philosophers (for instance, Abrol Fairweather and Carlos Montemayor, "Curiosity and Epistemic Achievement" in Ilhan Inan et al. (2018)). Indeed, the whole tradition initiated by Berlyne seems to be in good standing: it is not plausible that Silvia and the tradition are simply wrong about the whole issue.

Finally, a terminological note: what is here called curiosity is in the cognitive literature mostly called "interest" (Berlyne, Silvia). Silvia explains that the two are the same (Sect. "Interest and Curiosity" in the 2006 book, to which we shall be referring throughout the chapter):

> The interest–curiosity distinction may be based on the different uses of interest and curiosity in everyday speech. People often use *curiosity* to refer to events that have yet to happen or knowledge they have yet to acquire. *Interest*, in contrast, is often used to refer to ongoing events or to describe events in the past. People say they are curious to see a new movie, but they say they found the movie interesting after having seen it. Yet linguistic differences don't always reflect psychological differences—everyday language can lead science astray. The challenge for research is to identify real differences in appraisal structure, subjective qualities, and motivational consequences. To date, no evidence suggests differences between interest and curiosity, so equating them is justified. (2006: 191)

To repeat, he concludes quite clearly: "To date, no evidence suggests differences between interest and curiosity, so equating them is justified" (2006: 191). I shall sometimes write "interest-curiosity" to remind the reader of differences in usage between epistemological and cognitive psychological papers.

Here is then the preview. For warming up we start with Silvia's characterization of interest-curiosity as an emotion and ask whether it is compatible with virtue account. We then pass to the main tradition in cognitive science of curiosity, the theories of appraisal, taking Berlyne and Silvia as our paradigmatic authors. The central issue for them, in particular for Silvia, is the coping potential of possible objects-targets of curiosity. We shall translate "coping" into epistemological terms as "understanding" and offer a brief virtue-theoretical re-reading of the cognitive insights.

We conclude with epistemological consequences of the proposal, in particular to the perspectives of naturalization, returning to the two prospective kinds, descriptive-explanatory and normative. We briefly argue that the first is unproblematic for curiosity and appraisal theories as the naturalistic foundations, and point, very briefly, to the way in which the normative task could be addressed.

Curiosity: Emotion and Virtue

Let us first note the breadth of examples, of various areas and possible objects of curiosity, from music, art and politics to fictional texts. The research on the breadth (as contrasted with the depth) of curiosity has been developing for decades; perhaps the most quoted author is Marie D. Ainlie, who has developed questionnaires to test the two contrasting qualities.[3] The relevant texts may serve as a warning to philosophers that their usual narrow lists of examples might be misleading.

Next, we can pass to an apparent puzzle. The first chapter of Silvia's 2006 book defines interest-curiosity as being primarily an emotion, indeed a positive and active one.

Here are three crucial sentences from his brief concluding statement (I am skipping his references here):

> We began by considering some defining components of emotions, such as expressions and antecedents, coherence among responses, and subjective experience [...] Measures of subjective experience consistently point to a positive-and-active quality to feelings of interest [...], and interest clearly facilitates motivation and learning. [...] Taken as a whole, the evidence supports viewing interest as an emotion. (2006: 30)

[3] Ainley, M. D. (1987). "The Factor Structure of Curiosity Measures: Breadth and Depth of Interest Curiosity Styles", *Australian Journal of Psychology*, 39, 53–59.

The virtue epistemologist might wonder how compatible this taxonomy is with his or her take on the epistemic status of curiosity. If curiosity is a virtue (or a vice, for that matter), is this compatible with its being an emotion? The puzzle is only apparent. Fortunately for our project, as we noted in Chap. 6, in last the three decades, an important movement has developed, combining the study of virtue and the work on emotions. I hope Silvia's characterizations of interest-curiosity are completely compatible with the mainstream virtue epistemological developments.

Let me mention one more issue connected to the introductory moves in the theory. In spite of the impressive breadth of examples, neither Berlyne nor Silvia show any interest in bad or low curiosity, the "poly-pragmosyne" that Plutarch nor his followers were obsessed with. Silvia mentions: /S/picing up a boring text by adding interesting details might backfire (2006: 64), which is his only mention of possibly "lower" interests, and this is all.

My guess is that the interest in the issue has moved from purely theoretically motivated cognitive science research to the educationally motivated branches. My own preferred cognitive approach to the matter comes from Susan Engel's work on the development of curiosity in children. In Chapter 7 of her 2015 book titled "Gossip" she discusses people's tendency to gossip, which, of course, includes the wish to know things that are objects of gossip, and which can be, to a large extent, object or low or bad curiosity.

She notes that "psychologists have so often taken a dim view of our motivations for gossiping", but stresses fact that "the truth is our gossip serves several purposes" (2015: 130), some of which are positive and quite important. Gossip involves "cultural learning": you learned basic facts about your culture by listening to gossip:

> [W]hen children gossip, they acquire three kinds of information— knowledge about the target of their gossip (who pushed whom, who lives in a big house, who kissed a classmate, and who got punished), knowledge about what is condoned and frowned upon in their social group, and knowledge about the person with whom they are gossiping. Though perhaps only a biologist would leap at the chance to look through a microscope, and only a historian would give anything for access to original documents, every

> Tom, Dick, or Harry is eager for the vast knowledge available just by leaning over the fence to talk. (2015: 142)

This is perhaps the good place to mention the stable combination of curiosity with other emotions: we quoted Cicero's idea that in great men curiosity often goes together with the "desire to be the first" and noted that it seems to hold for some great minds in the history of science. Does such a combination make intrinsic curiosity less virtuous? Well, other virtues might show the same tendency to unite or ally with less virtuous motives; a courageous policeperson might also have a bit of vanity and be proud of her courageous exploits. How should we judge it? Well, some bad alliances to throw a shadow upon the original virtue, but I think that in the case of scientific curiosity the situation is far from dramatic: our future scientist is no saint, and cannot be required to practice motivational ascetism.

The Theory of Appraisal

The Awakening of Curiosity

Silvia notes that Berlyne (1960) argued that a family of variables determines whether or not something is potentially interesting. He adds that the he "named these *collative variables* because they involve comparing incoming information with existing knowledge, or comparing several regions of a differentiated stimulus field. Four collative variables—complexity, novelty, uncertainty, and conflict—received the most attention" (2006: 33). Silvia then notes: "/A/ core assumption shared by all appraisal theorists is that cognitive appraisals of events cause and constitute emotional experience" (2006: 55). Such theories "describe an emotion's appraisal structure through a set of elemental appraisal components". His examples include appraising events as relevant to a goal, evaluating resources for coping with an event, making attributions of causality and responsibility, judging an event's congruence with a motive or goal, and assessing whether an action falls short of personal and moral standards. Here is a fine example of his:

> Anger … involves (1) appraising an event as relevant to a goal; (2) appraising the event as incongruent with the goal; (3) judging a threat to one's social- or self-esteem; and (4) blaming someone for the threat. (2006: 56)

And most importantly, for him, "each emotion has a unique appraisal structure" (55).

Now, for curiosity, the crucial qualities in the appraisal are novelty of the phenomenon presented and subject's ability to cope with the phenomenon. Novelty is the *conditio sine qua non*. Ability to cope is more challenging. Silvia suggests that people anticipate whether they will be able to cope with the phenomenon: if the answer is no, curiosity will not arise. If it is yes, the curiosity is being born.

Just a remark on novelty, before we turn to the central topic, the ability to cope.

First problem: the superficiality of novelty. In his handout "Interest and Epistemic Goodness" (2011: once available at his website) Michael S. Brady stressed how superficial the requirement of novelty might be. He mentions that interest and importance diverge, and points out that "we tend to find old, expected, familiar things comfortable or enjoyable, but are interested in things which are unexpected, unfamiliar, mysterious, baffling" (Brady 2011: 2).[4]

---

[4] In his "Curiosity and Pleasure" in Ilhan Inan, et al. (eds), (2018), 183–196, he develops his idea in an interesting way. Since "curiosity varies with novelty and coping potential, but not with the importance or significance of subjects and questions" (2018: 193), we should not look for the source of value in the objects themselves. And he continues:

> The objects of our curiosity have nothing valuable in common, therefore, that could intrinsically merit or make appropriate the interest we take in them. Nevertheless, although the objects and questions that trigger our curiosity have nothing valuable in common, they are unified by the very fact that we are curious about them. This suggests that what is valuable isn't some distinctive and identifiable feature that the truth on subjects that trigger our curiosity has in common; instead, what is valuable is the relational state of being curious about some novel topic or understandable issue. It is the desire to know that unifies all instances of curiosity, just as it is the desire that a sensation be occurring that unifies all instances of pleasure. Moreover, just as there is no incompatibility between pleasure being relational and intrinsically valuable, there should be no obvious problem with thinking the same about curiosity: It is intrinsically valuable for us to be curious about the truth on some subject, even if we cannot identify any feature or quality of the relevant truth that would warrant the desire to know. (2018: 194)

I tend to agree with his conclusion, as visible from the chapter on value and curiosity.

Novelty is philosophically unproblematic. However, even if we agree that novelty is a superficial criterion, I think that other appraisal variables are less so—in particular the ability to cope. No wonder that Silvia appeals to it even in the context of the simple novelty check:

> There is wide agreement—among psychologists, at least—on the appraisal variables that generate interest. One of the central appraisals is of novelty: "whether or not an event is new, sudden, or unfamiliar. For interest, this novelty check includes whether people judge something as new, ambiguous, complex, obscure, uncertain, mysterious, contradictory, unexpected, or otherwise not understood." (2006: 57)

Philosophically, the most interesting component determining the appraisal is coping potential. Here is Silvia:

> *Coping potential* refers broadly to estimates of resources, power, abilities, and control in relation to an event. (Ibid.)
>
> The simplest appraisal structure of interest, then, involves two appraisal components: an appraisal of novelty, broadly defined; and an appraisal of one's coping potential in relation to comprehending the obscure event. (2006: 58)

Note the importance of understanding and comprehending. This will be our central topic in the next section.

The Central Issue: Coping Potential and Understanding

Now, what does coping consist in? Let me try to show that *understanding is the central component of mature coping*, so that "coping potential" here reduces to the potential for understanding. For less mature cognizers, coping might reduce to a kind of elementary proto-understanding, or even to a simple capacity for correct action in relation to the item that provokes the interest. Consider a child facing a potentially interesting configuration.

"What is this?!" would be a normal question. Or, the child might just try to manipulate the configuration: once it has figured out how it works, it will stop. Let me document my claim by pointing to typical examples Silvia uses as instances of coping:

> The events that people find interesting can probably be described thematically as events that are not understood but understandable. Like other emotions, the experience of interest may change or end depending on continuing appraisals of the situation. Interest may end through the reappraisal of either appraisal component. The person may eventually understand the event—upon reappraising the event's novelty (broadly construed), the person would no longer judge the event as new or complex. Similarly, reappraising coping potential will affect interest. An initially interesting movie, for example, can become uninteresting when the viewers feel unable to form a coherent understanding of the narrative. Conversely, a confusing text can become interesting if its hidden meaning is revealed. (58)

Here, coping is described literally as understanding! And the identification continues. Silvia notes that "experts in art and music prefer relatively complex images and melodies, whereas novices prefer relatively simple images and melodies" (2006: 58), and mentions experiments testing these preferences. Let me add more references to experiments, with apologies for lots of quotations:

> A second body of work, also in the study of aesthetics, examines the effects of meaningful information on emotional responses to art. Several experiments show that titles enhance positive emotional responses to art by making art more comprehensible. Providing titles for abstract paintings increases the viewer's appraised ability to understand the paintings (Russell & Milne, 1997). In turn, people enjoy the art more, especially when the titles promote elaborated representations (Millis, 2001). Providing extensive information about a painting, such as the artist's biography and the context of the work, has a large effect on understanding and on emotions (Russell, 2003). Taken together, these experiments show how emotional

> responses can be enhanced by increasing appraisals of coping potential. (2006: 59)[5]

Here is more:

> Past research thus fits an appraisal position on interest: variations in appraisals of ability to understand affect feelings of interest. But it is never surprising when past research is consistent with a new hypothesis. After all, past research preceded the hypothesis. Congruent evidence from past studies offers indirect support at best. Several new experiments were conducted to test the predictions of an appraisal perspective (Silvia, 2005c). In the first study, people provided self-reports of trait curiosity, openness to experience, and appraisals of their ability to understand complex and abstract art. People then viewed random polygons that ranged from simple (4 sides) to complex (160 sides). Their viewing set was manipulated: One group picked the "most interesting" polygon; the second group picked the "most enjoyable" polygon. Appraisals of ability to understand significantly predicted the complexity of the most interesting polygon. As people felt more able to understand complex art, they picked highly complex polygons as being the most interesting. Only appraisals of ability predicted polygon choice; individual differences in curiosity and openness to experience were unrelated, showing discriminant validity for appraisals. Consistent with the interest–enjoyment differences reviewed in chapter 1, the most interesting polygon was significantly more complex than the most enjoyable polygon. Moreover, appraisals of the ability to understand predicted the level of complexity that people found interesting but not the level that people found enjoyable, thus discriminating between the appraisal structures of different positive emotions. (Cf. Ellsworth and Smith 1988, 60–61)[6]

---

[5] The sources are:

Russell, P. A., & Milne, S. (1997). Meaningfulness and the hedonic value of paintings: Effects of titles. *Empirical Studies of the Arts*, 15, 61–73.

Russell, P. A. (2003). Effort after meaning and the hedonic value of paintings. *British Journal of Psychology*, 94, 99–110.

Millis, K. (2001). Making meaning brings pleasure: The influence of titles on aesthetic experience. *Emotion*, 1, 320–329.

[6] Ellsworth, P. C., & Smith, C. A. (1988). Shades of joy: Patterns of appraisal differentiating positive emotions. *Cognition and Emotion*, 2, 301–331.

> A third experiment illustrated the joint role of appraisals of novelty–complexity and coping potential. People viewed simple and complex pictures taken from books of modern visual art. For each picture, they gave ratings of interest and ability to understand the picture. As expected, interest depended on both complexity and coping potential. For simple pictures, ratings of ability to understand were unrelated to interest. For complex pictures, however, ability strongly predicted interest—interest increased as appraised ability increased. These relations remained after controlling for possible confounds, such as trait curiosity and positive affectivity. (61)

What do the examples from art suggest as to what coping is? In brief, in this context understanding seems to be the typical case of coping. If I am curious, and I want to know why X is the case, my goal is understanding, stretching from most ordinary and banal (Why are my neighbors quarreling? Why did Miss Ruritania cheat her boyfriend?), to the most elevated ones (Why do basic cosmological constants have the values they have?). It is the interest in causes, reasons, expressed by appropriate why-questions.[7] And Silvia returns to complexity, novelty, uncertainty and conflict as determiners of understandability:

> One of the first judgments in the appraisal sequence, according to Scherer (2001), is a *novelty check*—whether or not an event is new, sudden, or unfamiliar. For interest, this novelty check includes whether people judge something as new, ambiguous, complex, obscure, uncertain, mysterious, contradictory, unexpected, or otherwise not understood. (57)

Arguably, if curiosity is directed to the "new, ambiguous, complex, obscure, uncertain, mysterious, contradictory, unexpected, or otherwise not understood", its central goal is achieving understanding, rather than arriving at isolated items of knowledge. The leading question is: this is ambiguous (complex, obscure, mysterious, contradictory); how should one understand it? And novelty is in the vicinity. The proposal brings together propositional and objectual curiosity. I conjecture that most

[7] For a fine analysis of appropriateness: see Whitcomb (2018), "Some Epistemic Roles for Curiosity" in Ilhan Inan, et al. (eds), 217–237.

interesting cases in which a single proposition is not enough to satisfy one's curiosity are cases where curiosity demands understanding

Notice that this cognitive scientific view of curiosity fits well our characterization of curiosity as a motivating and organizing trait. It recognizes the valuable object by its characteristics that have to do with coping: if the coping potential is right, curiosity will be awakened. And coping, of course, appeals to capacities the person has, wakes them up and mobilizes them. For sophisticated varieties of curiosity we have the goal, understanding, in a psychological motivational role, exactly as our motivation view of curiosity would have it.

What about less sophisticated varieties of curiosity, to be found in children and animals? The cognitive terminology is not suitable for epistemologists, so I propose that we call the genus of this sort of curiosity "motivation to grasp". Next, we should distinguish lower and higher drives within the motivation to grasp, calling the lower ones simply "drive" and higher one "desire".

So, for lowest kinds of motivation, suitable for animals, one would have the drive to grasp, the lowest common denominator of all. When my dog Bobo is eagerly expecting my wife to arrive, and instead it is the unknown person appearing at the door, Bobo is puzzled. We might say that he has the drive to grasp the new, and unexpected situation. He wants to know what is going on.

A very small child might have a drive to understand, and a toddler, a clear desire to understand some analogous phenomenon.

Suzan Engel notes that this drive has been studied by first-rate cognitive psychologists Kagan and Piaget decades ago. First, Kagan:

> In his book *Surprise, Uncertainty, and Mental Structures*, Jerome Kagan argues that surprise shapes our mental life. […] We remember things (information as well as events) that rattle our sense of familiarity. Surprise not only etches things in our memory— it leads us […] to probe the source of surprise. We seek to understand what we didn't anticipate. Curiosity, in other words, can be understood as the human impulse to resolve uncertainty. (9)

Then, she passes to the earlier work, due to Piaget:

> Like Kagan, Jean Piaget thought humans were uniquely driven to make sense of their experiences (Piaget 1964b). It wasn't enough, in Piaget's view, to be able to navigate the world—humans, even four- month- old infants, are predisposed to understand the objects and events they encounter. This understanding emerges, in Piaget's framework, as a result of the young child's effort to explain the unexpected. Piaget thought that when young children confront an object or sequence of events that doesn't fit their mental schema, they attempt, however unconsciously, to understand why. (10)

And here is her summary of the achievements of both:

> Both Piaget and Kagan, like Berlyne, saw curiosity as a fundamental human urge. But they offered two essential additions to Berlyne's conception. First, their definitions stipulate that the internal urge is hitched to the outer world by way of thoughts concerning whatever event, information, or object an individual doesn't expect or understand. And second, both Kagan and Piaget viewed this powerful urge as the engine of early development. As the following pages will show, this second point is key. (10)

The important point for us is that the drives from the genus "motivation to grasp' are dovetailing with each other. The most elementary motivation to grasp and thus some to know (as in my Bobo) is the beginning of all; it then dovetails with the drive to understand, and this then dovetails with the mature desire to understand.

Let me add that this phenomenon speaks in favor of placing knowledge and understanding in the same category. If understanding is a particularly valuable species of knowledge, the growth in complexity, from grasping to understanding can be easily placed within a theory of human knowledge. If understanding is quite distinct from knowledge, we have a puzzle: how to bring together the phenomena that obviously fit each other.

The sketch of the drive to understand naturally points to the further question, namely why did nature give it to us at all. And there is literature available on the naturalistic origin of curiosity. Here I shall mention two connected interesting sources. Robert T. Pennock, in his excellent 2019 book on curiosity in science, dedicates the whole first chapter, "A Virtuous Instinct", to the omnipresence of curiosity in non-human animals. He ends up with the phenomenon very close to our "drive for understanding:

> To summarize, curiosity is a beneficial intralife disposition, leading organisms to follow up on deviations from the expected in the hope of discovering some new or improved generalization. Again, this will not be true everywhere for every kind of organism, but it is critical for organisms in environments that change on intralife time scales. Inflexible, hardwired behavioral patterns will not let an organism adjust to changing conditions. Even if following curious deviations from the norm does not lead to new resources, following up on differences may be useful for maintaining one's current resources; deviations from expectations may be a signal that the environment has changed, or that signs one has previously relied upon have changed, or even that one's sensors are malfunctioning. Which takes us back to confusion. (2019: 15)

He quotes as the main predecessor Charles Darwin and comments his statement in a letter that we have an "instinct for truth, or knowledge or discovery" and that "having such an is reason enough for scientific researches without any practical results ever ensuing from them" (2019: 4).[8]

We shall return to the topic of naturalization in a few lines. Before passing to the concluding section let me mention very briefly one further question, to which Silvia dedicates a chapter (Chap. 6).

Where does the continuing interest come from? I have been interested in languages since my early age, and this interest is distinct from a momentary curiosity in, say, Arabic words in the Bosnian dialects-language which I understand perfectly. His proposal is that the initial curiosity gives us initial emotional satisfaction (when I learned a bit of my first foreign language, I felt satisfied). We understand the causes and consequences of this emotional experience, we attribute the emotion production to them, and then we look for more emotional experience,

[8] He quotes as his source: Charles Darwin, "Letter to John Stephens Henslow", April 1, 1848. Darwin Correspondence Project, "Letter no. 1167": http://www.darwinproject.ac.uk/DCP-LETT-1167.

searching similar causes that will produce similar consequences. He calls the resulting proposal "emotion-attribution theory". Here is a quote:

> **The Development of Interests**
> If interests arise through attributions for emotional experience and the consequences of such attributions, then we can make predictions about how to manipulate the development and change of interests. First, changing emotional experience will change the development of interests. Creating positive emotional experiences is obviously relevant, but all of the emotions play important roles in the origination of interests. Whereas interest and enjoyment facilitate the development of interests, negative emotions inhibit interest development by changing people's expectations regarding an activity's emotional consequences. If people experience negative emotions during a formerly interesting activity, they will make attributions for these emotions. Should they attribute these emotions to the activity, their expectations about the emotional consequences of the activity will shift. Expectations of interest and enjoyment will seem less plausible, and expectations of negative emotions will seem more plausible. [...] Second, the development of interests can be influenced by manipulating attributions for an emotion. If ideas about the cause of an emotion affect what people expect and do, then changing these ideas will affect interests. (145)

I am not impressed with this theory of permanent curiosity. The emotional talk sounds deflationary. It might be correct, for example, like my curiosity about Arabic words, but it is not sufficient for impressive cases, like Galileo's abiding curiosity: "Oh, I enjoyed observing the moon so much, and I want to feel more enjoyment. So, let me look at Mars through my new telescope!" We can and should widen the scope of the "emotional" and look for other factors that will be more persuasive at least for these impressive cases.

Conclusion: Epistemological Consequences

We have noted that the main focus in cognitive understanding of curiosity is on the coping potential and understanding. The drive to understand and, in more sophisticated cases, desire to understand seem to be fundamental for human curiosity! If Silvia and the tradition are right, this is *the central message of cognitive psychology* concerning the epistemology of curiosity.

The task of descriptive naturalization thus seems to be within the power of virtue theorists and cognitive scientists. Epistemologically (or epistemically) important features of curiosity find their analogue, and probably their naturalistic grounding in the cognitive features of human dealing with novelty, complexity and other appealing features of external reality. Curiosity is in its highest form the desire to understand, or at least grasp in a way close to understanding. There are connections with understanding in science (and philosophy).

In the chapter on science, we mentioned that we shall discuss the importance of the desire to understand as the highest form of scientific curiosity.

And there is the naturalistic, descriptive grounding for this thesis, prominent in the literature in cognitive science of curiosity.

Here is a good place to say a few words about the natural way of descriptive-explanatory naturalization of curiosity. Abrol Fairweather rightly connects it to his "first bridge", connecting science and virtue epistemology, namely to the desire "to ground their epistemic psychology (person level cognitive dispositions and the processes that count as their manifestation) in current work in evolutionary biology and cognitive science" (2014: 3). But then he comes with a warning of "danger" in descriptive naturalization of the functioning of cognizing individuals. Here is his formulation:

> While this first bridge shows both prospects and problems for virtue epistemology on the empirical front, there might be a general concern that, even if the more optimistic lines win out, empirical work in the relevant sciences will replace authentic epistemic theorizing, and will essentially result in the end of virtue epistemology once the relevant scientific accounts have been fully worked out. This is the familiar worry about "imperialist" naturalism, or replacement naturalism. (2014: 3)[9]

[9] Fairweather tries to suggest that the danger is not great: science is important for virtues, but virtues are also important for science, which he describes as "second and third bridge" connecting the two; the second has to do with virtues helping to solve the indeterminacy problem, and the third with the role of virtues in history of science:

In the present case there is a clear answer: the epistemological interest typical for philosophers stays alive in the dialogue with cognitive explanation; indeed, the dialogue between the two seems to be potentially lively and useful to both sides. I see no dangerous prospect of cognitive account "replacing" the virtue-theoretical epistemological one, since the former is not deflationary, nor of the kind that would lead to the problematization of the latter. Epistemological consequences of descriptive naturalism seem to me welcome and not problematic.

What about normative naturalism? The appraisal theoreticians suggest that curiosity is epistemically successful, and that its success is due to its specific structure, revealed by their research on curiosity. It is driven by the desire to cope with new, conflicting and puzzling cases, and its persistence is to a large extent explained by its success. The centrality of understanding, or of understanding-achievements (in simpler cases, with animals and infants as the principal heroes), does a lot of explanatory job, and no mysteries are left to make the virtue theory normatively problematic.

So, we might conjecture that that norms leading to this success are the right means to the valuable end. This would then offer a beginning of a normative naturalization of the activities of human curiosity, and cognitive science would give us a basis for a workable naturalism, both descriptive and normative in this area. We shall return to these issues in the last chapter, in the section on naturalization of virtue theory of curiosity.

However, the second and third bridge discussed below suggest interesting ways in which the imperialist worry is actually less threatening for virtue epistemology because epistemic virtues are essential elements in the success of the sciences. We see this in recent work on virtue theoretic solutions to underdetermination, work on theory virtues and in specific discoveries in the history of science itself. This suggests a fruitful partnership, but now with virtue epistemology informing the epistemology of science, rather than scientific results informing virtue epistemology. (5)

References

Berlyne, Daniel. 1954. A Theory of Human Curiosity. *British Journal of Psychology* 45 (3): 180–191.

Berlyne, Daniel E. 1960. *Conflict, Arousal and Curiosity*. McGraw Hill.

———. 1966. Curiosity and Exploration. *Science, New Series* 153 (3731): 25–33.

Brady, M. S. 2011. Interest and Epistemic Goodness. Handout available at http://www.ou.edu/spring08conf/papers/Brad

Ellsworth, P. C., & Smith, C. A. (1988). Shades of joy: Patterns of appraisal differentiating pleasant emotions. *Cognition and Emotion*, 2(4), 301–331. https://doi.org/10.1080/02699938808412702

Engel, Susan L. 2015. *The Hungry Mind: The Origins of Curiosity in Childhood*. OUP.

Fairweather, Abrol, ed. 2014. *Virtue Epistemology Naturalized Bridges between Virtue Epistemology and Philosophy of Science*. Springer.

Inan, Ilhan, Lani Watson, Dennis Whitcomb, and Safiye Yiğit, eds. 2018. *The Moral Psychology of Curiosity*. Rowman & Littlefield International Ltd.

Kashdan, Todd B., and Paul J. Silvia. 2009. Curiosity and Interest: The Benefits of Thriving on Novelty and Challenge. In *Handbook of Positive Psychology*, ed. R. Snyder Charles and Shane J. Lopez, 367–374. Oxford University Press.

Pennock, Robert T. 2019. *An Instinct for Truth Curiosity and the Moral Character of Science*. MIT Press.

Silvia, Paul J. 2006. *Exploring the Psychology of Interest*. Oxford University Press.

Whitcomb, Dennis. 2018. Some Epistemic Roles for Curiosity. In *The Moral Psychology of Curiosity*, ed. Ilhan Inan, Lani Watson, Dennis Whitcomb, and Safiye Yiğit, 217–237. Rowman & Littlefield International Ltd.

9

The Curiosity of Science

Introduction

When curiosity turns to serious matters, it's called research.—Marie von Ebner-Eschenbach (*Aphorisms* 1994)

In this chapter we shall investigate the importance of curiosity in scientific research.[1] Let me motivate the investigation by quoting Albert Einstein; he saw curiosity as crucial for scientific inquiry:

> The important thing is not to stop questioning; curiosity has its own reason for existing. One cannot help but be in awe when contemplating the mysteries of eternity, of life, of the marvelous structure of reality. It is enough if one tries merely to comprehend a little of the mystery every day. The important thing is not to stop questioning; never lose a holy curiosity—Einstein (1955: 65)

[1] Big thanks go to the participants of the Philosophy of Science conference in Dubrovnik, in April 2019, and in particular to Dunja Jutronić, who pointed to a lot of defects in the original presentation, to Majda Trobok, for support and assistance, and to Marko Grba, for help with the history of physics.

N. Miščević, *Curiosity as an Epistemic Virtue*, Palgrave Innovations in Philosophy,
https://doi.org/10.1007/978-3-030-57103-0_9

But, how does curiosity lead to scientific inquiry? This should be a general topic in the epistemology of science. The first task would be classifying the kinds of scientific curiosity; such classification might turn out to be quite important for understanding scientific research. It would lead the researcher to compare theoretical curiosity, which leads to understanding, and practical curiosity, which directly motivates laboratory work, to investigate the ways the two interact, and the link between desire to understand and the desire to learn how to apply the understanding reached. Here, the similarities and differences between varieties of curiosity are crucial.

Indeed, curiosity has been explicitly linked to science throughout its whole history. For the historical use of the term for the interests of scientists, witness, for example, the name "Academia Naturae Curiosorum" (Academy of Those Curious about Nature; later known as the Leopoldina) established in the imperial city of Schweinfurt in 1652 by a handful of German physicians (Daston 2011).

Interestingly, some great researchers tend to neglect this fact. Here is an illustration. Steven Shapin, writing about scientific revolution in his 1996 book, asks what motivated it. He begins by referring to desire for knowledge and truth, and says that one might take as the answer the "[d]esire to produce and extend knowledge" (1996: 119). This sounds like an appeal to curiosity, but then he says that such answer, referring to "search for truth" is "plausibly general but too general". And he explains: "Just because that motive is plausibly general, it cannot effectively discriminate between strands of practices" (1996: 120). And this is all he has to say about the role of desire for knowledge. Quite consistently, his (and Shapin's) air-pump book (1985) also brings no material on this topic. One would expect Shapin to *distinguish between various species within the desire for knowledge as a wider genus*, and to show the role of each species (desire for knowledge of particular facts, of causal structures, etc.) in the complex motivation of the phenomenon he is studying. He didn't do it.

Two recent books on curiosity concentrate on its role in scientific revolution, Huff (2011) and Ball (2012). However, they don't explicitly distinguish types of curiosity, nor show how different kinds came to play different important roles. Huff's story is rather factual, Phillip Ball's erudite but anecdotal, with no theoretical underpinning. Similarly, a wonderful book by István Hargittai (2011), with the telling title *Drive and Curiosity: What Fuels the Passion for Science* (Amherst: Prometheus), promises to tell us more about what fuels science, but then says nothing of curiosity, and very little on drives. In short, there is lot of talk on the history of science, but no systematic attention to kinds of curiosity.

The new excellent book by Robert T. Pennock (2019), with the telling title *An Instinct for Truth Curiosity and the Moral Character of Science*, brings considerations of science into the context of curiosity as an epistemic virtue. He addresses several questions that we do not address here, and, on the other hand, talks very little about the topics like scientific revolution which we interest us. So, fortunately, his book is complementary to our investigations, and I recommend it to readers interested in philosophy of science, and further issues linked to scientific curiosity.

In this chapter, I shall concentrate upon the role of curiosity in scientific inquiry, taking Scientific Revolution as my favorite period. I hope to offer a useful classification of motives tied to the desire for truth or curiosity, and very, very briefly indicate or illustrate what kind of role each played. Here is then the preview. Our main topic will be varieties of scientific curiosity. We shall first look at motivation and distinguish intrinsic and extrinsic motives, placing curiosity in the first group. Then we concentrate upon it, distinguishing, high and low, as well as theoretical and practical curiosity. Within the first one we quickly pass over propositional versus objectual contrast, and focus upon the division between more isolated (atomistic) fact regarding curiosity and the understanding-directed curiosity, stressing the crucial importance of the latter.

We then pass to a brief summary of the history of scientific revolution, applying the curiosity theory to the history of science.

Our next topic is the division of cognitive labor, and the way the types of curiosity are socially organized in science. The final topic is more polemical: what about theories that claim that curiosity is not the driving force in science? Can we replace curiosity with matters like ignorance? We hope to offer a sketch of an answer to the central issue in epistemology of scientific curiosity: how does it lead our inquiry?

Motivation—Epistemic Versus Non-epistemic

Let us start with a few words about science driving motivation(s) in general. What we are interested in here is the motivation to come to know, either know that something is the case, of know how to do, or achieve something. We shall take both kinds of motivation as belonging to the "cognitive" (epistemic) domain and treat them as kinds of curiosity. Within this epistemic domain we can also distinguish intrinsic curiosity from instrumental curiosity Compare for instance the following two statements: first, I am curious if the gravitation causes this movement, I am curious about how to measure the air pressure *in order to* find out the weight of air. The first points to my intrinsic curiosity, the second to curiosity that is instrumental, in service of some other desire to know.

Individual motivation can be positive, neutral or negative, and the first two can be high or low. On the clearly positive side we have individual desire for self-perfection motivating one to proceed to self-cultivation; Matthew L. Jones describes its function in the scientific revolution in his 2006 book with the telling title: *The Good Life in the Scientific Revolution Descartes, Pascal, Leibniz, and the cultivation of virtue*.

On the lower side we have ambitions to win competition, which are ubiquitous in the life of science. Historians of science talk about "the rush to discover" and of "discovery of discovery". David Wootton in his history of scientific revolution (2015) points to "the "Columbus model": the discovery of America showed that there are things to be discovered, not mentioned in the available classics, and that a discovery can be "an unprecedented event" (2015: 55). He describes the process that resulted from it "the discovery of discovery" (in Chap. 3 of the 2015 book).

Nowadays, the importance of discovery is seen as *the* priority in research. More than half a century ago Robert Merton argued that the basic currency for scientific reward is recognition, especially recognition for being the first person to come up with an idea (I am following the summary of his work offered by P. Godfrey Smith (2003: 123). So much about the "lower" individual motivation. On the "high" general side the most discussed motivation is the control of nature.

Let us say more about both sides, first about a possible combination of high and low individual motivation, and then about the control-of-nature motivation, very prominent in philosophical literature, especially continental one. A very common type of combination combines intrinsic inquisitiveness with ambition—"lower" motivation. There is an early classical formulation concerning it, that we mentioned in Chap. 2: Cicero speaks of love of truth and "craving for precedence" (*appetitio principatus*). The inquiry of truth (*veri inquisitio atque investigatio*) is thus combined with the craving *impuls toward preeminence* (*De Officiis* Bk 1/13). This combination is indeed very often found in biographies of famous scientists. Since we shall concentrate on Scientific revolution, let me mention a story about Newton that sounds like a joke; here is a quote from biography written by Gale E. Christianson:

> In light of his interest in books and all things mechanical, one would have expected Isaac to do well in his studies, but that was not the case. Henry Stokes put him in the lowest form, or class, where he ranked second to last out of more than 80 students. Newton himself later admitted that he "continued very negligent" in his studies at King's School. Events took a radical turn one morning as Isaac was walking to class. The boy ranked just above him kicked him in the stomach, which provoked thoughts of revenge. When school was over for the day, Isaac challenged his classmate to a fight, and they went into the nearby churchyard to settle their differences. Though smaller than his opponent, Isaac fought with greater determination, beating him until he declared that he would fight no more. He then rubbed the defeated boy's face against the church wall for good measure. Still this was not enough. Isaac began applying himself to his studies and soon rose to become the top student in the school. (1996: 16)

Similar stories have been told about Niels Bohr,[2] Charles Darwin,[3] Thomas Alva Edison,[4] Werner Heisenberg[5] and to some extent Stephen Hawking.[6]

So much about the individual non-epistemic motivation. Let us now pass to high-level social not primarily epistemic motivation. The Frankfurt School, first Horkheimer and later Habermas, has been stressing non-cognitive interest in dominating nature. In his *Knowledge and human interests* (1971) Habermas famously distinguishes three interests. First, empirical sciences incorporate a technical quasi-cognitive interest; (in our terms, high curiosity-how, extrinsic to genuinely cognitive goal). Second,

[2] Here is a summary from the biography by Abraham Pais (1991):

> Bohr wanted the admiration of his brother: A central figure in Niels' life was his brother Harald, younger by a year and a half, who became an outstanding mathematician. … According to a school friend of both, 'The relationship between the two brothers was the most beautiful imaginable. Niels had other good friends during his boyhood but Harald was his only real friend and confidant. (I believe this was mutual.) They admired each other, Niels counted himself for nothing and Harald for everything—and vice versa.

[3] Here is Janet Browne on Darwin:

> Darwin's honour as a gentleman—as he understood it—was at stake (18). In the competitive scientific world in which he chose to live, publication, originality, and priority made a delicate trio. It was easy to succumb to the temptation to be secretive or be overly quick to publish; and yet the creator of any fresh insight, then as now, must eventually relinquish possession and place his or her ideas in the public domain in order to be given credit for advancing knowledge. The primary spur for Victorians like Darwin was not so much to gain individual power, as it might have been for the politician, nor wealth, as for the businessman, but reputation and professional pride—the need for recognition of the value of one's endeavour by others in the field. Browne (2002: 17)

[4] Gene Adair (1996). *Thomas Alva Edison: Inventing the Electric Age* (OUP), tells stories about bad reaction that acts of curiosity of the very young Alva (Edison) provoked in his parents:

> The scoldings and whippings that followed such episodes did not diminish Alva's curiosity. Full of questions, he was always ready to test whatever anyone told him. When he asked his mother why the goose sat on its eggs, her answer—"To hatch them"—sent him off to a neighbor's barn, where he curled up with some goose and hen eggs and tried to hatch them himself. (16)

[5] Helmut Rechenberg talks of Heisenberg competing with his older brother and finding in his own mathematical excellence the means of proving himself in front of his father (August. 2010: 10 ff).

[6] White, Michael, 2011 *Stephen Hawking: a life in science.*

"historical-hermeneutic" sciences incorporate a practical one ("practical" meaning moral and political). Finally, critically oriented social sciences incorporates the emancipatory cognitive interest (see, for example, 1971: 308).[7] Since we talk here about natural sciences, we should concentrate on the first interest. Habermas notes that sciences offer law-like hypotheses about nature, but then present this offering as being in the service of prediction. And then he claims that the meaning of such predictions is "their technical exploitability". The relevant interest is the one in technical control over objectified processes. I mostly disagree. The proposal assumes a broken tie between natural science and (objective) truth. It allows for no intrinsic scientific curiosity (search for truth for its own sake) present, at least not at socially important level.

So much about non-epistemic motivations of scientists. What have we gained by briefly discussing them? We have provided the natural context in which the epistemic motivation, the curiosity-driven efforts take place in scientific life. Now we can turn to these, and stay with them till the end of the chapter.

Curiosity: Practical Versus Theoretical

As regards theoretical curiosity we have at least the fact-directed one (curiosity whether, what and the like and the understanding directed one (asking—why, how possibly and the like). The more practical one is curiosity-how to φ.[8] The theoretical-practical contrast is well documented in the literature. For example, Ernan McMullin in his 2017 article in *Blackwell Companion to Phil of Science* recounts the history of dominant values in science, starting from the former and then coming to the latter. In Antiquity, the dominant goal for science, formulated by Aristotle, "was an understanding of the world around us in terms of its causes" (2017: 553); we also find the goal of truth, "understood as conformity between mind and world". But, at the end of seventeenth century a dif-

[7] Horkheimer, "Traditional & critical theory", p. 194 ff, originally "Traditionelle und kritische Theorie", 1937.

[8] As we noted before, motivation can in all cases be either intrinsic or extrinsic; if the latter, we should distinguish between individual goal and social goal and also prudential goal and moral goal.

ferent goal becomes prominent: science should tell us how to transform the world. This then extolled curiosity-how and gave it the pride of place.

The contrast, is of course, very important. Curiosity how become autonomous, and it gave birth to technological devices, and then laboratories; the further very important development along the same lines was the establishment of networks of laboratories.[9] Let me mention Nikola Tesla, famous for his imagination, which enabled him to visualize complex experimental procedures, involving sophisticated machinery. (See, for instance, Sean Patrick 2013). Another interesting case is Ciolkovsky, driven, by his own proclamation by intrinsic practical interest: how do a I build a missile (a rocket), was his obsession. And he was not very interested in actual application—his interest-how was intrinsic and non-instrumental.

The development of curiosity-how took two directions, internal and external. On the internal side, it, and then each of its sub-kinds, become relatively autonomous (in value-motivation and investigative activity). The most spectacular development is the curiosity-how explosion with

[9] Here is a nice illustration of the connection of curiosity-how, technology and discovery, from Toby Huff (2011); he does not offer any general principles about the role of kinds of curiosity that apply here. So, here is Huff on telescope as discovery machine. He talks about three contexts that are needed to understand the significance of the telescope:

> First, it is the emblematic instrument of the modern scientific revolution.

> Second, and most practically, the telescope transformed the practice of astronomy in the seventeenth century: it transformed astronomy from a plodding science into an active, exploratory inquiry that constantly looks for new discoveries. In that sense, the telescope was a newly invented discovery machine […] At the same time, the telescope is a portable laboratory that could be taken anywhere in the world and used to explore the heavens. (18)

> The third context concerns the telescope as a precision instrument. (19) (Huff, Toby E. (2011) *Intellectual curiosity and the scientific revolution: a global perspective*, Cambridge University Press)

development of laboratories: with them curiosity-how became institutionalized.[10]

On the other hand, interaction of more theoretical and more practical interest was crucial for the whole history of science. Let us very briefly mention one famous example: the birth of electrodynamics. Faraday's high-quality curiosity was first of all in satisfying curiosity-how (to perform an experiment, tease out a variable).[11] Faraday thus experimentally discovered interaction between magnetism and electricity. Then came Clark Maxwell, a high-level theoretician. His curiosity was ultimately curiosity-why, intrinsic, high, optimistic, active, long-term and going into depth.[12]

This suggests a general picture: the kinds of curiosity organize the cognitive faculties needed—an important point not noticed in the literature I have looked at. We shall be documenting this picture till the very end of this chapter.

---

[10] Here is Maxwell on the most famous case, the Cavendish Lab:

> Experiments of this class—those in which measurement of some kind is involved, are the proper work of a Physical Laboratory. In every experiment we have first to make our senses familiar with the phenomenon, but we must not stop here, we must find out which of its features are capable of measurement, and *what measurements are required in order to make a complete specification of the phenomenon*. We must then make these measurements, and deduce from them the result which we require to find. (Maxwell from *Cavendish Lab History*, p. 10)

[11] And here is where his curiosity-how lead him in his life:

> [T]hrough his experience as Davy's assistant Faraday learnt the many skills of an experimental chemist: blowing glass, constructing apparatus, accurate weighing of samples and titration, to name but a few. These skills remained with him throughout his career and were extended as he engaged new areas of science and developed new, but often impressively simple, pieces of equipment to interrogate nature. At the Royal Institution he had one of the best-equipped laboratories in Britain and he often spent his own money on purchasing new equipment. He was a brilliant experimentalist who effectively combined both head and hand in devising new apparatus and executing novel experiments. Throughout his life he maintained a keen interest in technical processes and machine design; he joined the Society of Arts in 1819 and later served as chairman of its chemical committee which vetted new designs and processes. (11) (Geoffrey Cantor, David Gooding, and Frank A. J. L. James 1991 *Faraday*, Macmillan)

[12] As we know from his biography. Basil Mahon (2004).

We can be more precise, however. When we talk about combined interests (= sub-kinds of curiosity) and the role of experiment—the most interesting role is the one of curiosity *how* in the service of *understanding*. The story of Higgs boson and CERN is a good example of curiosity-*how* in the service of *understanding-directed* curiosity.[13]

Similarly with crucial observation helping to decide between two or more understanding-offering theories.[14]

K. Knorr Cetina gives in her 1999 book an account of the changes induced in physics and biology by the increased sophistication of technology used for research. Here the technological means often dictate what abilities are required from the researcher, so that the organizing power of curiosity is mediated by the demands of technology. Her rich examples, prominently in Chap. 3 of the book, involve the development of collider, mostly at CERN, and the demands on researchers for its optimal use at various stages.

A book could and should be written about the relation of experiment-observation and desire to understand, but here we have to rest content with a few sentences.

It is clear that curiosity-how plays a crucial role in all the examples we mentioned. Normally, curiosity organizes the abilities the researcher has, but we can surmise that influence goes in both directions: if you have specific abilities, you will be attracted to issues that demand them, and once you chose the issue to deal with, the decision will influence the organization and activation of your capacities.

It is time to turn to theoretical curiosity.

---

[13] Roger Wolf 2015 *The Higgs Boson Discovery at the Large Hadron Collider* Springer. Krause, Michael, 2014 *CERN: how we found the Higgs boson*, World Scientific Publishing.

[14] A famous example of interaction of observation and understanding is offered by Einstein-Eddington's investigation of and the Sun and Mercury 1919. The eclipse that occurred on May 29, 1919, enabled Arthur Eddington to ascertain that the light rays from distant stars had been wrenched off their paths by the gravitational field of the sun. This affirmed the prediction of Einstein's theory of general relativity, ascribing gravity to a warp in the geometry of space-time, that gravity could bend light beams.

Theoretical Fact-Directed Curiosity Versus Desire-to-Understand

Let us first look very fast to the simplest kind of theoretical curiosity, the fact-directed one. It has been extolled in the empiricist tradition, starting at least with Bacon (as documented by Ball, 100 ff.).[15] The basic idea is simple: we need facts, and indeed a lot of facts before we even start with hypotheses and theories. We shall contrast this fact directed curiosity with understanding-directed variety that searches for connections, causes and big causal structures. In comparison to such understanding-directed variety the fact directed one is almost "atomic".

Let me illustrate this by reminding you of a famous experiment, Pascal's Puy-de-Dôme experiment 1648. His guiding question was simple: what is the pressure like at various heights? (see Wootton 2015). The mere list of data, correlating pressure with height, has been seen as a very important result. The research is not presented as understanding-directed, but as discovery of facts, although the general theory of pressure will easily yield understanding when connected to the data.

We have already noted the sociological fact, stressed by Merton, and studied by his followers till the present day, that the priority in discovery is "the basic currency for scientific reward"; some of the glory involved goes to factual discoveries, following the kind of inspiration exemplified in the glorification of Columbus's discovery.

Let me mention a complication however, to which I alluded shortly in the section on curiosity how. Consider the early discovery of radioactivity. One of the protagonists, Henri Becquerel describes the crucial experiment in a very simple way. He took a certain amount of uranium bisulfate, combined it with potassium, wrapped it in "Lumière photographic plate" (optical photographic plate) and a paper sheet to protect it from sunlight, and let it stay wrapped for a whole day. "Upon developing the photographic plate I recognized the silhouette of the phosphorescent substance in black on the negative" (212).[16] And that was it! Becquerel, by his own

[15] Although he also stressed the ideal of the "knowledge of causes", see Ball, 79.

[16] Henry Becquerel "Natural radioactivity", the 1896 paper originally just titled Comptes rendus, 122, reprinted in Shamos, Morris H. (ed.) (1959), *Great experiments in physics*, Dover.

description, concluded that the uranium bisulfate is emitting a new, hitherto unknown kind of radiation.

Now consider the skills needed to perform the experiment, basically the ability to pack a substance in a few sheets of paper of various kinds, to the skills needed to conduct sophisticated facts-establishing research in later work on radioactivity. Of course, the fact-directed curiosity lies behind the process, but what skills are to be mobilized is co-determined by available technologies. The mobilizing force of curiosity is thus acting quite indirectly, co-determined by technologies of increased sophistication, very differently from the functioning or ordinary curiosity. But again, we can note how a given kind of curiosity leads research and indirectly mobilizes intellectual and physical capacities of investigators.

Since the prioritizing of facts has been so omnipresent in the logical positivist tradition, we shall not spend more time on it; we shall stop here and return to its role in Scientific revolution below. Let me conclude by quoting the opposite intuition, stressing the importance of theorizing as opposed to fact collecting. It stems from the woman who discovered pulsars, Jocelyn Bell-Burnell:

> So we were very, very lucky in this physics teacher who was very good and with a subject like physics once you understand it it's that easy. *It's the one, I think, that involves the least learning. You don't have to learn lots and lots and lots of facts; you just learn a few key things, and if you really got hold on them, then you can apply and build and develop from those. So I think it appeals to brains with rather few cells* (Italics NM). He was a really good teacher and showed me, actually, how easy physics was, and I think that was the key. Otherwise I must have gone some other way.

We now pass to understanding-directed curiosity, the highest kind of theoretical curiosity that plays crucial role in motivating scientific efforts. As Einstein once wrote, probably having in mind isolated factual knowledge: "Any fool can know. The point is to understand."[17]

[17] I was not able to identify the source. It is mentioned on the web on many place, for instance, https://brightdrops.com/albert-einstein-quotes

We shall take understanding to involve seeing connections, in particular the causal ones, getting knowledge-why.[18] Understanding in general is a species of knowledge; not merely factual knowledge that Einstein was probably denigrating, but a more holistic, connections-focused knowledge.[19] The theoretical understanding-directed curiosity is exemplified by questions like "how do these phenomena hang together?" and, in the simplest case just "why such-and-such happens"? The practical interest is quite powerful: in many cases, once you know why, you can influence the process, you get the power of influencing nature, influencing other people by acting on right variables provided by your understanding. Epistemic value of understanding derives from its richness, cognitive relevance and role in manipulating causes. No wonder that understanding is central for social organization of knowledge, and therefore for social epistemology of inquisitiveness.

The only author who has explicitly raised the question of understanding-driven curiosity is Jonathan Kvanvig (2003, 2012); he has done it in general, but not in relation to science in particular. Here is his list of values tied to understanding-directed curiosity:

> [T]o have mastered such explanatory relationships is valuable not only because it involves the finding of new truths but also because finding such relationships organizes and systematizes our thinking on a subject matter. (2003: 202)
>
> Such organization is pragmatically useful because it allows us to reason from one bit of information to other related information that is useful as a basis for action, where unorganized thinking provides no such basis for inference. Moreover, such organized elements of thought provide intrinsically satisfying closure to the process of inquiry, yielding a sense or feeling of completeness to our grasp of a particular subject matter. (Ibid.)[20]

---

[18] Here I agree with Stephen Grimm, and thank him for inspiring discussion. For him the object of understanding is dependency relations, which we can mentally represent by means of causal maps.

[19] Here we disagree with important authors, like Pritchard, Elgin, Zagzebski, Kvanvig and Wayne Riggs. The disagreement has little consequence for philosophy of science, so we shall not discuss it here.

[20] He stresses the "objectual" understanding, not of isolated statements but of a larger domain as a whole (169), the typical experience involves objectual understanding of some topic or issue or

I agree with Kvanvig about the importance of understanding-directed curiosity, and think it also holds for scientific curiosity.

Where does this understanding-directed curiosity come from? The answer is not easy to find in literature. There is a strange situation about the research on motivation to understand in science. The positivist tradition has relegated understanding to a completely subordinate position, but a lot of work in the opposite direction is being is being done recently, by authors like Stephen R. Grimm, Christoph Baumberger and Sabine Ammon Henk W. De Regt M. Stuart (see References). But there is almost nothing about what in general motivates scientists to understand! (Individual biographies tell interesting stories about particular motives of particular scientists, but no systematization.)

Here is one exception, which I can present in a very sketchy manner, because of lack of space. Cognitive psychologist Alison Gopnik stresses the importance "aha" feeling, associated with an explanation achieved. Her title is telling: "Explanation as Orgasm":

> It all started with "Aha" feeling, associated with an explanation achieved
>
> My hypothesis is that explanation is to theory formation as orgasm is to reproduction—the phenomenological mark of the fulfillment of an evolutionarily determined drive. From our phenomenological point of view, it may seem to us that we construct and use theories to achieve explanation or that we have sex to achieve orgasm. From an evolutionary point of view, however, the relation is reversed, we experience orgasms and explanations to ensure that we make babies and theories. (Gopnik 2000, 300Gopnik Gopnik). (Explanation as orgasm and the drive for causal knowledge: The function, evolution, and phenomenology of the theory formation system. In Explanation and cognition, edited by F. C. Keil and R. A. Wilson, Cambridge, MA: MIT Press. (299–324)

Peter Lipton offered a friendly comment, connecting the hypothesis to the question of scientific curiosity:

phenomenon (170) and objectual understanding is the central goal of the phenomenon of curiosity (170) (2012, Curiosity and a Response-Dependent Account of the Value of Understanding", in Timothy Henning & David Schweikard (eds.), Knowledge, Virtue, and Action: Putting Epistemic Virtues to Work, Routledge, 151–174).

> First, whatever one thinks about evolutionary psychology, it is plausible to say that the desire for this pleasant sensation serves to motivate the search for understanding. Scientific understanding [...] has instrumental value. Knowing about causal history, about necessity in nature, about how things might have come about, and about how to work one's way around a theory may all make us better able to cope with our environment. But the payoffs are often highly indirect and only in the long term. So it is just as well that there is also such an immediate and personal phenomenological payoff to keep us going. [...] The third point is that there is a natural though substantial extension of Gopnik's hypothesis concerning the causal role of the "aha" feeling that I would like to promote. Her point is that the feeling motivates the activity of hypothesis construction. My suggested addition is that the feeling may also play a role in guiding that activity. (53, Peter Lipton, "Understanding without explanation understanding", in Henk W. de Regt, Sabina Leonelli, and Kai Eigner. *Scientific understanding: philosophical perspectives*, 43–63)[21]

In literature on understanding in science, this is what one finds.

Scientific Revolution—From Facts to Understanding

Scientific revolution is usually praised as victory of curiosity. Two recent books on curiosity, Huff (2011) and Ball (2012) concentrate on its role in scientific revolution. They are extremely rich and informative. But, as we mentioned in the Introduction, they don't distinguish types of curiosity, at least not explicitly, nor show how different kinds came to play

[21] Let me also quote Stephen R. Grimm on the "aha" experience:

> [T]here is something that is distinctively like to "see" or "grasp" (or, at least, to seem to see or to seem to grasp) why things are one way rather than another. Whether one prefers to describe this phenomenology in terms of something as dramatic as the "aha" experience, or whether instead one prefers to think of the phenomenology in more subdued terms, the basic idea is that at least some phenomenology accompanies the exercise of our understanding system, a system that is apparently well connected with how things stand in the world. (89, "Reliability and the Sense of Understanding", in Henk W. de Regt, Sabina Leonelli, and Kai Eigner, *Scientific Understanding: Philosophical Perspectives*, 83–99)

different important roles. Here, we shall go the opposite way, starting from information collected in Ball and Huff, pointing to a few (very few) well known aspects of the history and briefly point out how they can be systematized with the help of typology of scientific curiosity (and curiosity in general), but let me start from an overview.

First, non-epistemic versus epistemic motivation. In a previous section we talked about non-epistemic motivation, and mentioned Gale E. Christianson's (1996) wonderful description of the combination of the two in the life of young Newton. One could go through the whole history of scientific revolution and investigate the motivations of each of the protagonists.

Here, we shall stay with epistemic motivation and kinds of curiosity. We mentioned Pascal and his deep interest in factual measurements, exemplified by the Puy-de-Dôme inquiry. However, the factual curiosity had a much wider sway. Ball talks of "omnivorous curiosity" (2012: 128) of the Royal Society. He discusses the ambitions of Robert Hooke, Boyle's pupil and the curator (since 1662) of the Royal Society, manifesting absolute priority of factual curiosity, but no order in questions asked.[22]

Here is Hooke's list of topics to be covered, reproduced by Ball (2012: 125):

- Theory of motion
  - of Light
  - of Gravity
  - of Magneticks
  - of Gunpowder
  - of the Heavens
- Improve shipping
  - —watches
  - —Opticks
  - —Engines for trade—
  - Engines for carriage

Inquiry into the figures of Bodys

- —qualitys of Bodys

[22] Ball notes that rationalists on the continent, above all French (Descartes, Mersenne, Pascal, Leibniz), were less fact-crazy.

And he notes the prevalence of interest in facts, the Facts first norm:

> Bacon's stricture that theories should be deferred until all the facts were at hand sometimes became, with the Royal Society, an apparent determination to avoid causal explanations altogether. Instead the Fellows were apt to collect facts much as some collectors and antiquaries stocked their shelves with curiosities. (2012: 132)[23]

But what about normal scientific interest in causes and understanding? Ball notes very informatively:

> No amount of self-discipline could entirely suppress the instinct to ask "why?" Instead, the Fellows tried to justify their reticence with humility, or to phrase hypotheses so tentatively that it was not entirely clear they had been advanced at all. To the issue of interpretation, Thomas Sprat averred, "the Society approaches with as much circumspection, and modesty, as human counsels are capable of: They have been cautious, to shun the overweening dogmatizing on causes on the one hand; and not to fall into a Speculative Scepticism on the other." (2012: 133)

Now, if we pass from the beginnings to the culminating period of scientific revolution, we shall notice the dramatic change of the leading type of curiosity: it is now the understanding-directed curiosity that runs the show. With Newton it is questions like Why do planets go around the sun? that take the center stage. The anecdote about apple falling points to

[23] Hooke's list of inventions he should make goes a bit beyond the factual curiosity, but only a bit:

- A Way of Regulating all sorts of Watches or Timekeepers
- The true Mathematical and Mechanical form of all manner of Arches for Building
- The true Theory of Elasticity or Springiness
- A new sort of Object-Glasses for Telescopes and Microscopes
- A new Selenoscope [for surveying the moon]
- A new sort of Horizontall Sayls for a Mill
- A new way of Post-Charriott for travelling far
- A new sort of Philosophical-Scales

the width of understanding searched for: planets behave like apples, although this is not obvious at the first glance.

Unfortunately, the apple anecdote is a myth. It took decades to Newton to offer a unified account, not a moment of inspiration (as documented by I. Bernard Cohen and George E. Smith 2004: 5). The proposed understanding of understanding, that identifies it with seeing connections, in particular the causal ones shows here its teeth. Here is the non-anecdotal material, about Newton finding the common pattern for the behavior of projectiles, satellites and planets:

> A projectile, if it was not for the force of gravity, would not deviate towards the earth, but would go off from it in a right line, and that with a uniform motion, if the resistance of the air was taken away. (I. Bernard Cohen and George E. Smith, *The Cambridge Companion to Newton*, Cambridge University Press, 2004)

And this points to the crucial hypothesis, as formulated by Newton himself:

> Introduction 1–31 apple p. 5 And by increasing the velocity, we may at pleasure increase the distance to which it might be projected, and diminish the curvature of the line which it might describe, till at last it should fall at the distance of 10, 30, or 90 degrees, or even *might go quite round the whole earth* before it falls; or lastly, so *that it might never fall to the earth, but go forwards into the celestial spaces*, and proceed in its motion in infinitum. And after the same manner that a projectile, by the force of gravity, may be made to revolve in an orbit, and go round the whole earth, the moon also, either by the force of gravity, if it is endued with gravity, or by any other force, that impels it towards the earth, may be continually drawn aside towards the earth, out of the rectilinear way which by its innate force it would pursue; and would be made to revolve in the orbit which it now describes; nor could the moon without some such force be retained in its 4 orbit. (Bk 1, DEFINITION V)[24]

[24] For a lot of further unifications see Harper's chapter in *Cambridge Companion to Newton.*

Famously, Newton also developed the required mathematics. The interaction of geometry and mechanics was, of course, already exercising its powerful influence, above all with Johannes Kepler and the New Astronomy. Here is Newton's explanation of what he was doing:

> In mathematics we are to investigate the quantities of forces with their proportions consequent upon any conditions supposed; then, when we enter upon physics, we compare those proportions with the phenomena of Nature, that we may know what conditions of those forces answer to the several kinds of attractive bodies. And this preparation being made, we argue more safely concerning the physical species, causes, and proportions of the forces. (Bk I, Prop. LXIX, scholium, p. 192)

Mathematics is important, since evidence of patterns increases understanding, since it makes possible patterns of dependence *evident*. In the same way it makes the *connection* between the rectilinear and curvilinear motion clear and evident (what "conditions of forces answer to several kinds of attractive bodies"). Mathematics enables axiomatizing the mechanics, and axiomatization dramatically increases the surveyability of patterns (thanks, Nenad Smokrović!), and thus their understandability. We might conjecture that such *evidence of patterns of connection is responsible for increased understanding*. (See also Tappenden, *Proof Style and Understanding*.)

Our modest examples, from Pascal through the Royal Society to Newton, suggest that historians of scientific revolution should pay more attention to kinds of curiosity prominent in its history, and probably helping to shape this history! This concludes our main line of presentation, the story of types of curiosity and their role in the history of science.

Of course, much more remains to be said. For example, Karin Knorr Cetina gives an account of the development of understanding of scientific instruments in the twentieth-century science (1999: 58). Here, the desire to understand is turned toward the means, and the results are quite impressive, but we shall not be retelling Cetina's story here.

Instead, we now turn to two further issues we mentioned in Sect. 1. The first is the division of cognitive labor, and the way the types of curiosity are socially organized in science. The second issue is more polemical:

what about theories that claim that curiosity is not the driving force in science? Can we replace curiosity with matters like ignorance?

Division of Cognitive Labor and Social Epistemology of Curiosity

Until now, we have mostly talked about individual researchers, in particular great names like Newton or Faraday. In recent years, however, there is a growing interest in social epistemology, and it has also been turned toward curiosity. Let us say just a few words about it. Fred D'Agostino gives a fine rationale describing the first and immediate object to be studied:

> [A] *project of enquiry might be sub-divided into different tasks that could then be distributed* across different individuals or teams within the community. But that, of course, is what's required for a genuine division of labor in a cognitive or epistemic context. ("From the Organization to the Division of Cognitive Labor, *Politics, Philosophy and Economics* 8(1): 101–129 (2009) (web))[25]

Some decades ago Thomas Kuhn has pointed to the division in his "Objectivity, Value Judgment and Theory Choice" (Kuhn 1977), and he raised the question of whether there was a tension between individual and collective rationality in science. Philip Kitcher (1990, 1993) picked up on this question. He envisioned a stylized scenario of some set S of N scientists, each looking to pick between different projects in the set R to uncover some significant truth:

> Suppose that a scientist is dedicated to a particular inquiry: the scientist's overriding concern is to bring this inquiry to a conclusion by discovering the true answer to a particular question. In terms that I shall employ more systematically in later sections, this imagined scientist is a pure epistemic

[25] See also Ryan Muldoon 2017–2018, Diversity, Rationality, and the Division of Cognitive Labor' in Boyer-Kassem, Mayo-Wilson, &Weisberg. *Scientific Collaboration and Collective Knowledge.* Oxford University Press, 78–93.

agent, one for whom the primary goal is to reach an epistemically valuable state. (*The Advancement of Science—Science Without Legend, Objectivity Without Illusions*, 1993: 308)

Let me summarize the way I see social division of labor. I would distinguish more cooperative patterns of division from non-cooperative ones. On the side of the latter we would have competition and vigilance. Vigilance involves constant checking of the info received; in bad cases it might lead to wholesale rejection, either just or injust.[26] Competition normally leads to stratification: if group A is competing with group B the most reasonable strategy is further to divide the labor. First, to distribute the task in accordance with (presumed) capacities of various members of the group, second, to enhance the efficiency thanks to distribution, and finally to balance the results into harmonious whole. This distribute-enhance-balance pattern is to my mind responsible for many important pluses and minuses of the social division of epistemic labor. On the positive side we have the act of offering to each participant the place optimal for her in the larger whole, which goes in favor of the dignity of each researcher. The resulting enhancement might be the second big plus. For instance, suppose that the motivation is understanding-directed curiosity: why such-and-such happens. Once you understand why, you know which variables influence which. Then, you search the way to act upon the relevant variables to achieve the right effect. Application then verifies understanding and you have a virtuous circle.

But there are minuses: for instance, loss of general picture at all levels except the highest. The next problem is with division of labor as such: if you are not among the best, you are assigned the less skill-demanding task. And there is the famous political problem: less privileged members of the group, like women and others doing work that is (assumed to be) less skilled. One natural question concerns the possible danger of a downward spiral.

Karin Knorr Cetina has offered a fine analysis of division of labor in two areas, two "epistemic cultures" as she calls it, high energy physics and

[26] The work of Elisabeth Fricker on social patterns of unjust rejection, and epistemic injustice in general, is the standard work in this domain.

biology, in the course of twentieth century (1999, Chaps. 7 and 8). The circle of division and enhancement is much more obvious in the first, she claims. It creates a real "communitarian mechanism", with massively common publications, increasing anonymity of individual researchers ("erasure of the individual as epistemic subject", as she calls it (1999: 166), and completely "distributed cognition" (1999: 175).

The general question is quite clear in the face of these developments and changes of patterns of scientific curiosity. Consider the good aspects of the division, epistemic, moral, prudential, political (justice-related): can we preserve and strengthen them, and at the same time minimize, or even better, eliminate the bad ones?

So, the social organization of scientific curiosity and ways to satisfy it is a huge, recently opened ground for research, with interesting theoretical and moral-political implications. But here, I have to leave if for another occasion.

Curiosity and Its Competitors

We have argued for the central place of curiosity in scientific effort: it often motivates the effort, and often mobilizes and organizes the abilities and means of doing science. But is curiosity really unique in this regard? Aren't there competitors around that can do the job equally well or even better? In the last two decades there has been an effort to invent and introduce competitors to curiosity for the roles we mentioned. The most prominent is ignorance; others, like attention are too close to curiosity itself to present a serious threat. So we have to look at curiosity and ignorance; does ignorance drive science rather than curiosity? Yes, answers Stuart Firestein in his very successful 2012 *Ignorance: How It Drives Science*. In a more systematic way Jens Haas and Katja Maria Vogt (2015) offer their notion of "investigative ignorance" as the force driving scientific research.

Let us first look briefly at Firestein. In the first part of his book he speaks freely of curiosity-driven research: "like Darwin and his worms, the biologist's curiosity is enough for him or her to spend a lifetime mastering the details of another creature's life history" (2012: 62). But why doesn't he stay with curiosity then? He explicitly says that he "prefers curiosity-driven research to hypothesis-driven research", and then notes that "unguided curiosity is often not a helpful source of research questions". But is it really different from curiosity? Or is it just a sub-species (as the adjective "guided" suggests? Here is his reason for avoiding the term:

> Although you might have thought that curiosity was a good thing, the term is more commonly used in a derogatory manner, as if simple curiosity was too childish a thing to drive a serious research project. (2012: 79)

Of course, "ignorance" also has bad connotations, so the decision about the word to use has to be made on other grounds. And here, it is the issue of motivating force that is decisive. To use and autobiographical example, like Firestein does all the time, I am totally ignorant about football, the most popular, perhaps the only popular sport in my country, Croatia. But I am also perfectly unmotivated to learn about it; my ignorance does not drive me to do anything.

This is the essential point: ignorance itself, without curiosity does not motivate. Firestein himself starts the chapter that immediately follows his rejection of term "curiosity" by raising "the question of how you can use ignorance to understand that activity broadly called Science" (2012: 82). And then he notes, as the crucial matter, that "scientists love questions". But loving question goes much farther than just being ignorant; it crucially involves being curious. Would ignorance without curiosity drive research? Clearly not. I conclude that ignorance, as described by Firestein, is not competitor "to loving question" and being curious.

We now turn to Jens Haas and Katja Maria Vogt and their notion of investigative ignorance.[27] Here is what they say. They start with an epistemological framework, summarize in two statements:

A: Ignorance is the absence of knowledge. One either knows something or is ignorant of it.

[27] I would like to thank Katja Vogt for a wonderful discussion I had with her in 2019 in Rijeka.

B: Cognizers hold a range of doxastic attitudes: suspension of judgment, beliefs with higher and lesser credence, hypotheses, suppositions, postulates, and so on. (2015: 19)

Next, they propose their idea of "investigative ignorance":

> Consider next the doxastic attitudes that figure in inquiry. In an effort to avoid dogmatism, inquiring minds are committed to assessing—with significant and at times seemingly excessive effort—whether they are in a position to make claims about the world. They assign probabilities to assumptions, qualify views as preliminary, formulate hypotheses, and so on, adopting the attitudes of Investigative Ignorance. (Ibid.)

They note that "it is not enough to say that a cognizer is ignorant in the absence of knowledge (A). Something also needs to be said about the doxastic attitude she holds (B)" (2015: 19).[28] And then they say about inquirers: "They come across something they are ignorant of and that they would like to know. Hence they investigate" (2015: 20).[29]

But doesn't this bring us back to curiosity? If you "come across something" X you are ignorant of and that you would like to know, aren't you then curious about X? The word "investigative" in the phrase "investigative ignorance" points to investigativeness, which is just an active subkind of curiosity.

Let me briefly mention Lorraine Daston (2000, 2001), who offered "attentiveness" as her competitor to curiosity. She integrates curiosity into her wide notion of attention, herself pointing to wondering and curiosity (*Staunen und Neugier* in the original German) (2000: 14). No wonder that in the historical sections she refers to "cabinets of curiosity", as typical examples.

---

[28] And they make reference to Plato's *Meno* at this point.

[29] And they add:

> (B) supplies a description of where they stand in their quest for knowledge: as of now, one merely has a hypothesis, a promising model, inconclusive evidence, and so on. (2015: 20)

So, I conclude that ignorance is no competitor to curiosity.[30] If taken *simpliciter*, as in Firestein, it does not explain what motivates the scientist, and if taken as "investigative ignorance" it coincides with investigativeness, and thus reduces to curiosity. Similarly with a second potential competitor, "attention" proposed by Lorraine Daston.

Conclusion: Curiosity—Motivating and Organizing Epistemic Force in Science

The central idea of the chapter is that curiosity is a motivating and to a large extent organizing epistemic virtue in science. Therefore, the study of curiosity and its kinds can help can help systematize historical research concerning science and understand the cognitive structure of scientific inquiry (theoretical, applied, hybrid, "high", "low" etc.).

We took as our example, the cognitive structure of scientific revolution (with its heroes, individual scientists and "societies"). It is typically not addressed in this way. For instance, the two extant studies of the history of science, with "curiosity" in their title, the one by Ball and the other by Huff don't go beyond collected facts. The attention to scientific curiosity can help systematize the rich relations between theoretical and applied science, and their motivational underpinnings.

These branching sub-kinds and their results enter a web of cooperation either with means-end organization, or simply as parallel interactive activity. And curiosity, individual or group, or wider social, dictates the desired structure of cooperation and organization. These processes are accompanied by and partly made possible by social deep division of labor.

A spectacular testimony of all these facts is the history of scientific revolution. Revolution itself was to a large extent generated by scientists' curiosity, and its development followed the main lines of division of kinds of curiosity. It combined curiosity how, with a lot of experimental work, and more theoretical curiosity, and within the latter, it passed the path from predominantly factual curiosity to the global desire-to-understand.

[30] Compare Inan (2016).

Desire to understand, Darwin's "keen pleasure in understanding any complex subject or thing", is arguably the highest form of scientific curiosity. I would assume that epistemic value of scientific understanding derives from its richness, its organized and systematic character. the latter enhances its cognitive relevance and enables the scientist to connect local discovery to its distant and up to the present moment unexplained consequences. Finally, it plays the central role in manipulating causes, thus steering technological application.

Again, in the case of the search for understanding, it is curiosity that mobilizes, steers and to significant extent organizes scientific efforts—a fact never really analyzed in the literature, nor used in understanding the history of scientific research.

In short, Einstein was right in talking about "the holy curiosity of inquiry".

References

Adair, Gene. 1996. *Thomas Alva Edison: Inventing the Electric Age*. Oxford University Press.

Ball, Philip. 2012. *Curiosity: How Science Became Interested in Everything*. The University of Chicago Press.

Basil, Mahon. 2004. The Man Who Changed Everything: The Life of James Clerk Maxwell, Willey.

Browne, Elizabeth Janet. 2002. *Darwin Power of Place*. Vol. 2. New York: Knopf.

Christianson, Gale E. 1996. *Isaac Newton and the Scientific Revolution*. Oxford University Press.

Cohen, Bernard I., and George E. Smith. 2004. *The Cambridge Companion to Newton*. Cambridge University Press.

Daston, Lorraine. 2000. Preternatural Philosophy. In *Biographies of Scientific Objects*, ed. Lorraine Daston. University of Chicago Press.

———. 2001. *Eine kurze Geschichte der wissenschaftlichen Aufmerksamkeit*. München: C.F. von Siemens Stiftung.

———. 2011. The Empire of Observation, 1600–1800. In *Histories of Scientific Observation*, ed. L. Daston and E. Lunbeck. Chicago: The University of Chicago Press.

Einstein, Albert. 1955. Old Man's Advice to Youth: 'Never Lose a Holy Curiosity.' *LIFE Magazine*, 2 May. From the memoirs of William Miller, an editor, quoted in Life Magazine.

Firestein, Stuart. 2012. *Ignorance: How It Drives Science*. Oxford: Oxford University Press.

Godfrey, Smith. 2003. Theory and Reality: An Introduction to the Philosophy of Science, University of Chicago Press.

Gopnik, Allison. 2000. Explanation as Orgasm and the Drive for Causal Knowledge: The Function, Evolution, and Phenomenology of the Theory Formation System. In *Explanation and Cognition*, ed. Frank C. Keil and R.A. Wilson, 299–324. Cambridge, MA: MIT Press.

Haas, Jens, and Katja M. Vogt. 2015. Ignorance and Investigation. In *Routledge International Handbook of Ignorance Studies*, ed. M. Gross and L. McGoey, 17–25. London: Routledge.

Habermas, Jűrgen. 1971. *Knowledge and Human Interests*. Boston: Beacon Press.

Hargittai, Istvan. 2011. *Drive and Curiosity: What Fuels the Passion for Science*. Amherst: Prometheus Books.

Huff, Toby E. 2011. *Intellectual Curiosity and the Scientific Revolution: A Global Perspective*. Cambridge University Press.

Inan, Ilhan. 2016. Curiosity and Ignorance. *Croatian Journal of Philosophy* XVI (48): 285–304.

Kitcher, Peter. 1990. The Division of Cognitive Labor. *Journal of Philosophy* 87 (1): 5–22.

Kitcher, P. 1993. *The Advancement of Science*. Oxford: Oxford University Press.

Knorr, Cetina K. 1999. *Epistemic Cultures How the Sciences Make Knowledge*. Cambridge, MA: Harvard University Press.

Kuhn, Thomas. 1977. The Essential Tension, University of Chicago Press.

Kvanvig, Jonathan L. 2003. *The Value of Knowledge and the Pursuit of Understanding*. Cambridge University Press.

———. 2012. Curiosity and a Response-Dependent Account of the Value of Understanding. In *Knowledge, Virtue, and Action: Putting Epistemic Virtues to Work*, ed. T. Henning and D. Schweikard, 151–164. London: Routledge.

Marie von Ebner-Eschenbach. 1994. *Aphorismen*. Riverside, CA: Ariadne.

McMullin. 2017. Values in science, in W. H. Newton-Smith (ed.) A Companion to the Philosophy of Science, Blackwell, 550–560.

Pais, Abraham. 1991. *Niels Bohr*. Oxford: Clarendon Press.

Patrick, Sean. 2013. *Nikola Tesla: Imagination and the Man That Invented the 20th Century*. Clearwater: Oculus Publishers.

Pennock, Robert T. 2019. *An Instinct for Truth Curiosity and the Moral Character of Science*. MIT Press.
Rechenberg, Helmut. 2010. *Werner Heisenberg–Die Sprache derAtome Leben und Wirken—Eine wissenschaftliche Biographie Die "FröhlicheWissenschaft" (Jugend bis Nobelpreis)*. Springer.
Shapin, Steven. 1996. *Scientific Revolution*. Chicago: University of Chicago Press.
Wootton, David. 2015. *The Invention of Science: A New History of the Scientific Revolution*. New York: Harper.

10

Self-inquisitiveness: The Structure and Role of an Epistemic Virtue

Introduction

The chapter is concerned with self-knowledge and self-inquisitiveness, the cognitive-epistemic interest in oneself. I want to preserve the link to "curiosity", since the term is used in the wider epistemological context, but I want to avoid the strange sounding "self-curiosity".[1] So, I shall be using "self-inquisitiveness" rather than "curiosity about oneself", since the latter sounds like bad English, and "curiosity about oneself" is quite a mouthful. And I shall simply stipulate here that the two are synonymous.

We shall also very briefly address the connection between self-inquisitiveness, the resulting self-knowledge and wisdom. *Dao de jing* teaches us: "Knowing others is intelligence; knowing yourself is true wisdom" (Stanza 33).[2] Socrates would probably agree, and so would many of his followers. But why is self-knowledge important? Well, I should know myself in order to be able to predict how various outcomes will affect me, how I will probably react in various circumstances. We are often very keen on self-evaluation, both in purely cognitive issues (How reliable am

[1] Thanks go to the reviewer for pressing me on this point.

[2] Translation by S. Mitchell, Penguin (2018, Lao Tzu 2006).

N. Miščević, *Curiosity as an Epistemic Virtue*, Palgrave Innovations in Philosophy,
https://doi.org/10.1007/978-3-030-57103-0_10

I? How often am I wrong?) and in action-related cognitive ones (How good am I at deliberating? How often do I make prudential mistakes, and how bad are they? How often am I morally wrong?). One thus tries to get an insight into one's causal dispositions, active and passive. Within ancient view(s) of self-knowledge this causal-dispositional information was seen as a means of care for oneself: "Knowing what we are, we shall know how to take care of ourselves, and if we are ignorant we shall not know" (Plato: *Alcibiades 129a*, translated by Jowett). The practical importance of "self-critical perspective", crucial for ethics and views of human happiness and welfare geared to the care for oneself, was the basic motivational factor, not mere curiosity about states of the self, which become primary with Descartes, when care for oneself started being abandoned by philosophers.

Interestingly, the connection between self-knowledge and self-inquisitiveness has not been much discussed. Qassim Cassam is an interesting relatively recent example: in his excellent 2014 book he is very keen on the instrumental value of self-knowledge, but not keen on its intrinsic value nor on self-inquisitiveness.[3] I shall be arguing first for the relatively non-demanding view, that self-inquisitiveness is the motivating epistemic virtue for arriving at self-knowledge, and then for a more substantial view, that it is the basic epistemic virtue that organizes other epistemic virtues, both intellectual abilities and epistemic character virtues needed to arrive at important pieces of self-knowledge. The background of the argument is the general view that one could call "the motivating virtue account" according to which curiosity or inquisitiveness is the motivating and organizing epistemic virtue. We are inquisitive, curious and alert partly because of practical interests, searching the means for practical ends, partly because of pure need to know, or from both at the same time. Our curiosity is thus either pure or practical or mixed.

We can agree that a human being devoid of curiosity (in this wide sense) would have little motivation to arrive to true belief and knowledge. On the usual view of motivating virtues, this would seem to make it a virtue; since it is the main spring of motivation, we should take it as the motivating epistemic virtue. After all, wanting to know whether *p* gives

[3] For a discussion of Cassam see Miščević (2017).

the cognizer's particular instances of *p* as particular goals and knowledge and truth as the general epistemic goal. So, we have a truth-focused motivating virtue: inquisitiveness-curiosity having as its general goal reliable arriving at truth. This is, I have been claiming in favor of the motivating virtue account, the core motivating epistemic virtue. There is a multitude of questions of all sorts that we ask, whether, why, when, how, and inquisitiveness-curiosity caters to all of them.[4] Here, I want to apply this battery of questions to self-inquisitiveness, which has a long standing in the virtue-tradition at least since Socrates. In a perceptive comment Anthony Hatzimoysis has observed that "for the ancients self-knowledge is primarily a good to be achieved, whereas for the moderns it is mainly a puzzle to be resolved" (2011: 1). In a more recent comment, Ursula Renz (in her introduction to the 2017) book writes: "Considering the larger history, the importance assigned to the concept was just as often derived from the role self-knowledge presumably plays in the achievement of wisdom. Some philosophers even claimed that the acquisition of self-knowledge is the very end of philosophical inquiry; to engage in philosophy, they thought, is to explore and thereby to ennoble the self" (Renz 2017: 3). And she joins Hatzimoysis in deploring the fact that "moral or wisdom-related aspect of self-knowledge, which was quite important for the history of the concept, is largely absent in contemporary discussion" (ibid.). Authors writing about the meaning of life, like Nozick, have continued the tradition. Nozick starts his book on *Examined life* by noting that we "mostly tend [...] to live on automatic pilot, following through the views of ourselves and the aims we acquired early, with only minor adjustment (1989: 11). And he notes the loss we incur by neglecting self-examination. However, virtue epistemologists have stayed away from the issue of self-inquisitiveness. I want to propose that we turn to the topic, and indeed from the standpoint of mainstream virtue epistemology, and I hope that we can learn from connecting the two. I shall not be doing conceptual analysis of self-knowledge, but inquire into conditions and possibilities of it.

Here is then the preview. In the next section I briefly list some main varieties of self-knowledge and propose that the proper object of

[4] See Miščević (2016).

self-inquisitiveness is substantial self-knowledge, of items concerning one's character, dispositions and abilities-competences. For illustration, let me list two self-describing sentences from a psychological questionnaire, "I can say that I am a happy person", and "I am an anxious kind of person" (Ulrich Wiesmann et al. 2013: 12). It is not easy to achieve self-knowledge of one's dispositions. In Sect. 3 I sketch a picture of the virtue epistemology of self-knowledge, looking at two kinds of ignorance that have to be overcome in order to arrive at self-knowledge. The first is ignorance about character traits and temperament; I shall be taking the character of Othello as the paradigm for self-ignorance in these crucially important matters. Then I very briefly turn to ignorance about one's cognitive capacities, and appeal to the work of social and cognitive psychologists. In both areas I hope to document the importance and the role of skills, but also point to the fact that skills are not enough; what is also needed are epistemic character virtues.

The second part of the chapter finally turns to self-inquisitiveness and its role. Section 4 is the central piece of the chapter, and it argues for the fundamental role of self-inquisitiveness in arriving at self-knowledge, its ability to motivate, deploy and organize other epistemic virtues, directing them to the final goal. Section 5 goes further, but in a tentative fashion. It addresses the issue of the intrinsic value of self-knowledge, briefly discusses one recent proposal about value of self-knowledge in general, due to Michael Cholbi (web), and then tentatively proposes a response-dependentist account. Self-inquisitiveness is a deep trait of our nature, and *it* bestows intrinsic, non-instrumental value upon items of self-knowledge.

The concluding section summarizes the main proposals and connects them with the general picture of the role of curiosity-inquisitiveness in promoting knowledge, and with equally tentative proposal from Chap. 7 that the intrinsic value of knowledge in general derives from facts about human curiosity. If the direction is right, self-knowledge and self-inquisitiveness are paradigm examples of the rich roles of epistemic virtue in human cognition that deserve much more attention in the field of virtue epistemology than they have enjoyed in recent analytical efforts.

It also points to the Socratic teaching about wisdom essentially involving self-examination, and therefore self-inquisitiveness. Take Socrates' famous claim that "unexamined life is not worth living" (Apology, 38a). The examined life is a life of knowing oneself, recommended by the oracle, which seems to recommend to us a life of permanent self-inquiry, and attitude of permanent self-inquisitiveness. On the most popular reading, the value of examined life and of knowing oneself is of a high instrumental nature: this knowledge will make one morally more capable, prudentially more successful, and the like. But, in another famous place Apology offers a slightly wider picture; here Socrates is asking the Athenians whether they are not ashamed for caring only for "having as much money as possible, and reputation, and honor", but not caring nor giving thought to "prudence"[5] and truth, and how your soul will be the best possible?' (Apology, 29 d8-e3).

A few lines before (at 29b) he suggests that "the most reprehensible form of ignorance is that of thinking one knows what one does not know". One might read these passages as suggesting that knowledge and truth do have some value in themselves; in the given context that would imply that knowing truth about oneself is clearly intrinsically valuable.[6]

A much more radical, but in some respects limited line in favor of the intrinsic value of knowing oneself is present in the Aristotelian tradition, with the ideal of the Intellect thinking (about) oneself; it is not a kind of self-knowledge we would think of today, but it is worth mentioning.

Levels and Kinds of Self-knowledge

Let us start with a general picture of self-knowledge. One can talk about it and related attitudes (like justifiedly believing something about oneself) in two senses, propositional and objectual. The latter involves learning about some "internal" objects, for instance an emotion one has, being able to recognize it, connect it to other mental states, and the like. The former involves grasping of some truth about oneself, say knowing, or

[5] "Phronesis", which Grube translates as "wisdom".

[6] On the point of caring about truth (of one's beliefs) see the summary of proposed readings in Christopher Rowe's 2011 chapter on self-examination in *The Cambridge Companion to Socrates*. On the general topic of self-examination see also Richard Kraut (2006), "The Examined Life" in Sara Ahbel-Rappe and Rachana Kamtekar (eds.), *Blackwell Companion to Socrates*, Blackwell.

coming close to know the truth of some self-related proposition σ (for instance, coming to know (or firmly believe that I believe that *p*); I shall use σ (mnemonics for "self-") as the name of self-related content. (For more on propositional-objectual contrast see Chap. 3.) Here we shall concentrate on propositional inquisitiveness, but the results, I hope, will be applicable to the objectual variety as well.

As noted in the introductory section, I shall follow a further division between kinds of self-knowledge. The first kind concerns character traits, and other lasting dispositions to act and react. To continue with list of self-describing sentences from the psychological questionnaire mentioned, "I resign easily", and "I am satisfied" (Wiesmann et al. 2013: 12).[7] Such knowledge is usually contrasted with knowledge of occurrent episodes, like knowing that one is in pain. I shall shorten "causal and dispositional" to "CD" and talk about CD properties, self-attached CD knowledge. Central to the CD level is the causal structure of one's self. One causes things, acting in the world, and also, and crucially important for wisdom and care for the self, acting on oneself. Self-attached knowledge of one's CD level items concerns causally oriented active and passive dispositions of one's self; the ways of being (possibly) affected by various courses of things, and of reacting to them.

Roughly the same division has been drawn by Q. Cassam (2014: 29 ff.) and described by him as one between "trivial" and "substantial" self-knowledge. He writes: "Intuitively, knowing that you believe it is raining is a relatively trivial and boring piece of self-knowledge" (2014: 9). On the opposite, substantial side he places knowledge of one's character, values, abilities, aptitudes and emotions (2014: 29). He stresses the fact that acquiring such kind of knowledge is not direct and requires "a degree of cognitive effort". (I note a slight difference with Cassam: I would count all the dispositions, e.g., preference for some kind of ice-cream, as belonging to the dispositional group, and he would not.) The division also roughly corresponds to what Annalisa Coliva (2016: 51, 69) describes as a contrast between first-person and third person self-knowledge. Note that substantial self-knowledge goes well beyond phenomena of

---

[7] I am quoting these examples in order to point out that psychologists are as interested in self-attached CD knowledge as philosophers are; see below for more.

self-consciousness. To illustrate, Thomas Natsoulas, in his 2015 book *The Conceptual Representation of Consciousness*, lists six meanings, and at least four concepts of consciousness. He talks about the awareness meaning, the inner-awareness meaning, the unitive meaning and the general-state meaning (each mentioned in the title of a chapter), which all stay within first person (or trivial) variety, and none of them comes close to any kind of self-attached CD knowledge.

There are more detailed divisions, of self-knowledge, of course, for example one due to Robert Stalnaker (2008); we shall not need it here, however (I discuss it in Miščević (2017)). Why focus upon self-attached CD knowledge? We have at least three kinds of reasons for this. The first is the traditional philosophical interest: making sense of classical views of self-knowledge.[8] The second is the practical importance of "self-critical perspective", crucial for ethics and views of human happiness and welfare. Next is the interest of psychology as science, and the importance of interface between science and philosophy. Cognitive and developmental scientists are, of course, very interested in people's knowledge about one's dispositions and abilities. In the developmental psychology, the topic appears within the area of "self-concept development"; see, for instance, Michael Ferrari and Robert J. Sternberg (eds.) (1998). In the book the collaborators note that very early infants can "*begin to experience joy/pride about a result attributed to oneself* (italics in the original) (1998: 358). The agent, they note, has to have a capacity "to form single representations of self's agent or of production of stable outcome" (ibid.). But the thought that one is capable of producing a stable outcome concerns the causal-dispositional structure of infant's self, and clearly point to elementary CD knowledge. We have been quoting the questionnaire on "health-related Self-concept"; the authors discussing it, Wiesmann et al. (2013: 7–20), are interested in such CD properties as health-protective dispositions, health-protective motivation, vulnerability, health-risky habits and extrinsic/avoidant motivation (2013: 10).

Another area in which the research on what we have called CD knowledge concerns people's capacity for self-regulation. For example, successful self-regulation presupposes some knowledge of one's "self-efficacy", as

[8] See, for instance, Kamtekar (2017) and C. Moore (2015).

psychologists call it: "Self-efficacy beliefs refer to specific behaviors and arise from direct and indirect experience with the behavior. Although self-efficacy beliefs are behavior specific, the likelihood of developing strong self-efficacy beliefs can be attributed, in part, to stable traits" (see Hoyle 2010: 13).

We now turn to our main topic: the genealogy of self-attached CD knowledge, and the role of epistemic virtues, in particular self-inquisitiveness, in it.

The Virtue Epistemology of Self-knowledge—A Sketch

We shall assume here, as we do in the rest of the book, a wide framework of epistemic virtues, and count as such both abilities ("virtues" in the ancient Greek sense, revived by authors like Sosa (2001, 2007, 2015)) and epistemic character virtues, like self-inquisitiveness, and epistemic modesty.[9]

Let me start with virtues-abilities, specific to self-knowledge. Traditionally, introspection had the place of honor, but in a last few decades a new line of thought about sources of self-knowledge has emerged, thanks to the pioneering efforts of Gareth Evans; we can call them transparency abilities. They concern one's capacity to check one's beliefs and desires (and other propositional attitudes) by asking external questions. Suppose the relevant issue concerns my belief that *p*: do I have such a belief? I check it by asking whether *p* is the case; if my answer is yes, I believe that p. Generally, I pass from questions about *p* to the related piece of self-knowledge or something similar, like justified true belief about my propositional attitudes:

> [I]n making a self-ascription of belief, one's eyes are, so to speak, or occasionally literally, directed outward—upon the world. If someone asks me "Do you think there is going to be a third world war?", I must attend, in answering him, to precisely the same outward phenomena as I would

[9] Discussed at length in Roberts and Wood (2007) and J. Baehr (2011).

attend to if I were answering the question "Will there be a third world war?" (Evans 1982: 225)

Of course, this does not mean that the two questions are literally the same. The point is that the answer to the external, factual one is a certain guide to the answer to the internal, doxastic one. Call the abilities needed for taking the relevant step from the external to the internal "transparency abilities". I leave it open precisely which abilities are needed; Jordi Fernandez in Chapter 2 of his 2013 book proposes an undemanding account that I like a lot, "the bypass model", according to which the self-attribution is based directly on the basis for the first-order belief.

To pass to somewhat deeper CD properties, one needs general epistemic abilities, for instance inductive abilities, needed for understanding one's habits and reactions. To stay with the psychological questionnaire about health (from Wiesmann et al. 2013), items like "In the past, I often practiced unhealthy behaviors", "My lifestyle is risky" and "I find unhealthy behaviors enjoyable" are clear example of induction from memory data.

The examples bring us, however, to the other big group of virtues, to epistemic character virtues, in particular, specific abilities having to do with self-knowledge. To offer a personal example from the same genre, it took me a long time to understand that my lack of physical exercise is in a way an unhealthy behavioral style; I had enough of information and reasoning abilities, but I needed help from my friends to realize it, and indeed I did it very late in my life. I lacked courage to face my deficiency, and perhaps a dose of epistemic modesty-humility about myself. Other epistemic virtues listed in Roberts and Wood (2007) are firmness, caution, epistemic humility, autonomy, generosity and practical wisdom (see also Baehr 2011). We can add healthy curiosity about oneself that we have been referring to as "self-inquisitiveness".

So, we can think of one easy kind of topics of interest for self-inquisitiveness: momentary phenomenal states that are given to us with immediacy, although not infallible. ("Do I feel a pain in my neck?" would be a typical example.)

Next come causal-dispositional (CD) properties. Start with the easier sub-species, beliefs, desires and other propositional attitudes closer to conscientious surface: Do I believe that Trump will last 4 years as president? My favorite account of the ways to arrive at them is the transparency account mentioned above.

Finally, we have deeper causal-dispositional properties: Some of them concern one's character, temperament and standing dispositions to action: Am I prone to jealousy? Am I lazy by nature? Or, to quote examples from health questionnaire", I resign easily", "I am an anxious kind of person" "I am a vulnerable type of person" and"My lifestyle in risky". Others concern one's qualities, including cognitive ones: how good am I at spatial orientation? Am I talented for music? How good am I for math?

Let me concentrate upon CD knowledge of one's character, a clear sub-species of self-attached CD knowledge. We need some examples of people who find arriving at such knowledge difficult, or simply ignore its importance. Let me start by and taking a famous example from fiction, Othello, who brings misfortune to himself by not reflecting about his jealous and angry feelings. Late I shall pass to examples from cognitive psychology that concentrate about people's misjudgments about their cognitive and moral CD properties.

Othello thinks of himself as of someone who is not prone to jealousy:

> [O]ne not easily jealous, but being wrought
> Perplex'd in the extreme.

He becomes more self-critical when it's too late and all is done: "O fool! fool! fool!", he complains about himself. Othello seems to be aware of this, after the tragedy:

> Speak of me as I am;
> … then must you speak
> Of one that loved not wisely but too well;
> Of one not easily jealous, but being wrought
> Perplex'd in the extreme; of one whose hand,
> Like the base Indian, threw a pearl away
> Richer than all his tribe;

Now, imagine a more reflective and self-inquisitive Othello-R, who manages to arrive at a happy end in his marriage and his relations with friends and military colleagues.

He reflects about his dispositions. What does he need and, by assumption, has? How does he proceed? A simple case would be induction from introspective and memory beliefs. Start from the bottom and follow the arrow:

AM I 'EASILY JEALOUS'? YES.
↑INDUCTIVE REASONING
DO I FEEL JEALOUSY NOW? YES
DID I FEEL IT YESTERDAY? YES
...AND EARLIER? YES

What can motivate our Othello-R? Self-inquisitiveness, the one of ordinary instrumental variety, namely worrying about one's rather obvious habits:

I AM PRONE TO WILD, UNCONTROLLABLE ANGER
("RASH AS FIRE")
↑INDUCTIVE REASONING
I GOT TERRIBLY ANGRY YESTERDAY
AND DAY BEFORE YESTERDAY
AND DAY...

Othello-R can then easily put together his two lines of reasoning:

I AM PRONE TO EXCESSES OF ANGRY JEALOUSY
↑REASONING, etc.

PRONE TO JEALOUSY	PRONE TO ANGER
↑REASONING, etc.	↑REASONING, etc.
JEALOUSY FEELING 1	ANGER 1
JEALOUSY FEELING 2	ANGER 2

And he can proceed with further particular episodes, specific morals from them, and then a more general summary. This would give us the simplest picture, still pointing to a hierarchy of levels and of skills:

THE MORE OBJECTIVE CD KNOWLEDGE
↑REASONING, etc.
SINGULAR EPISODES OF SUBJECTIVE, 1st PERSON KNOWING

So far, epistemic virtues-skills are playing the main role. But they are not enough since self-awareness and inductive skills are impotent if the character is blocking them. Othello needs character epistemic or intellectual virtues, like humility, and we may assume that Othello-R has them. Indeed some practical wisdom might help him.[10] He would thus become aware that he loved "not too wisely", and would have a chance to correct his defect. We shall discuss another, and crucially important virtue that he badly needs, namely self-inquisitiveness, in the next section.

So much for Othello, and his ignorance of his emotional temperamental traits. We turn now briefly to the work of psychologists investigating people's ignorance and their misjudgments about their own cognitive abilities. They offer some hard data about the fact that philosophers often mention, but rarely document, namely that attaining substantial self-knowledge is difficult. We shall take as our compass the work of David Dunning, in particular his 2005 book on *Self-Insight*, and 2011 chapter on "the Dunning–Kruger effect". But we shall mention other authors from the same tradition. So, here is Dunning, with his mixture of pessimism and optimism:

> Ignorance makes a habit of sly and artful invisibility. But, perhaps, once we know of the trick, we become a little bit wiser in how to look out for and deal with this mischievous, significant, and hopefully not-too-frequent companion. (2011: 290)

So, how does sly and artful invisibility of self-ignorance function? In his book, Dunning gives hundreds of examples, and I shall limit myself to several kinds of phenomena. For instance, he speaks of what he calls "Barnum effect", namely the finding that people "swallow" general, non-specific descriptions presumably of themselves with no hesitation. He reminds us that Barnum was exploiting this weakness in order to impress

[10] For epistemic humility see Roberts and Wood, p. 237 ff.

people with his "mind-reading" abilities: you volunteer, come to the scene and he tells you a quite generic story, presumably about yourself. You end up impressed with his reading or you mind (Dunning 2005: 90).

Does knowledge of what particular traits are like help? No, he suggests: "if you ask people explicitly to provide concrete definitions of traits, the definitions you are likely to hear are ones that flatter the person you are asking" (2005: 101). The self-flattering also plays a role in self-ignorance about ones morally relevant traits and properties (Chapter 7 of the 2005 book).

What about knowledge of one's competences, a clear sub-species of self-attached CD knowledge. Dunning has, together with his student Justin Kruger discovered another worrying fact about people's knowledge, or rather ignorance, of this kind. The Dunning–Kruger Effect, as the object of the discovery came to be called, is a pattern of dramatic overestimation of one's competences by the least competent among relevant individuals. If you know some area moderately or well, you also judge the degree of your knowledge more correctly; but *if you are quite ignorant, you will judge yourself quite knowledgeable* (ibid.: 67). The effect is quite stable. The less you know about an area, the more knowledge of it you self-ascribe.

Worse of all: feedback fails to inform, and this for various reasons: first, it is normally probabilistic, so you can manipulate probabilities in your favor. Normally, it is incomplete, so you complete it in the way that suits you (2005: 67). It is often ambiguous, and sometimes biased (Othello was getting such a feedback from Iago, and the result is famous, or rather infamous).[11]

Dunning also proposes some remedies for this kind of weaknesses. For instance, he notes that "forcing people to consider information about the strengths, plans, and weaknesses of others causes them to curb their unrealistic optimism, albeit only a little)" (91).

Other weaknesses analyzed by Dunning are more intellectual. He talks about vagueness and ambiguity in traits under consideration. "I am a studious person." Yes, you do study, but how much? And how much is sufficient for being studious? Here, precision is the healing virtue.

[11] Compare papers from Alicke and Dunning (2005), especially in Parts 2 and 3 of the collection.

Other authors working in the same tradition confirm what has been said until now. A recent 2017 book by psychologist Tasha Eurich has a telling title *Insight: Why We're Not as Self-Aware as We Think, and How Seeing Ourselves Clearly Helps Us Succeed at Work and in Life*. Her suggestion is that one crucially needs feedback from others, and the willingness and ability to recognize its value (2017: 212). What is the right epistemic character virtue for this? Humility might be important, but what she describes is more like disciplined trust, a proper credulity disposition.

Let me finally mention another kind of difficulties for achieving self-knowledge, namely the broadly social pressures, for example, influence of gender, racial and class stereotypes that can, consciously of not, disfigure one's picture of oneself. They are now attracting attention of epistemologists in analytic tradition; see, for example the volume on *Implicit Bias and Philosophy* (M. Brownstein and J. Saul 2016) and also a fine paper on the topic "Implicit Bias and Qualiefs" by M. Fürst (to appear); the author argues for the existence of such deep social constraints that many people are completely unaware of.

To conclude, we were discussing relatively simple examples of the role of epistemic virtues in the achievement of self-knowledge, but I believe that the picture sketched is valid in general and that it can accommodate any level of complexity.

Self-inquisitiveness—The Motivating Virtue Account

Self-inquisitiveness, Intrinsic and Instrumental

What virtue can put one on the right track, when it comes to weaknesses just listed? We have seen that our reflective Othello-R would need ordinary intellectual abilities and some epistemic character virtues, like humility, that would help him to see his deficiencies, some autonomy, in relation to Iago and his testimonies, and some other in order to arrive to a happy end? Now we may ask what can motivate him and awaken all these abilities.

Given that he has no problem with his professional military and leadership capacities he can put them on one side and turn elsewhere. Indeed, some curiosity about his own emotional set-up might be the only important particular sub-variety needed. He, or his self-inquisitive counterpart Othello-R, might note that he felt very jealous day before the given day, and day before that, and again day before, and ask himself: wait, am I being constantly jealous? Desdemona behaves in the same way she behaved for years. Is it something about myself that pushes me to jealousy? And, in good circumstances, that might be a crucial step forward. What is it with me that makes me torture myself all the time? (A bit of a proper credulity disposition might help: I should not talk only to Iago, and trust him only, but ask my other mates how they see my bad mood, and heed their advice.) So, we can add self-inquisitiveness to the simple schema proposed above:

I AM PRONE TO EXCESSES OF ANGRY JEALOUSY
↑REASONING, etc.

I AM PRONE TO JEALOUSY	I AM PRONE TO ANGER
↑REASONING, etc.	↑REASONING, etc.
REMEMBER JEALOUSY FEELING 1	REMEMBER ANGER 1
REMEMBER JEALOUSY FEELING 2	REMEMBER ANGER 2

↖↗

SELF-INQUISITIVENESS
PARTICULAR, INSTRUMENTAL

Let us start from the bottom line. Othello-R is tortured by his suspicion concerning Desdemona. What is happening to me, why am I so restless, he asks himself. His attention moves away from Desdemona, and turns to himself. This might evoke in him memories of other episodes of jealousy (two penultimate lines on the left). A normal cognizer, and we assume that Othello-R is one, would generalize the morals of these memories rather quickly—this is the third line from the bottom on the left, pointing to reasoning. The result is a true generalization about oneself: I am prone to jealousy.

Pass now to the right side. Memory will also remind Othello-R of his proneness to anger: particular memories, inductive generalization and the general conclusion about oneself. Two kinds of capacities are needed, memory and inductive reasoning ability.

Now, Othello-R is in position to make the decisive step, to put the two together. I got angry-and-jealous to a high degree many times. "Now, that I am reminded of it, this seems to me a bit excessive." And we have the generalization, the top line. The generalization might turn out to be helpful; it moves the focus away from the object of jealousy, to its subject.

Indeed, we can see how self-inquisitiveness would mobilize the epistemic virtues needed: introspective abilities, memory, and reasoning abilities, and might encourage the modesty and proper credulity, if Othello-R has them.

We can thus add the decisive bottom line, preceding the memory activity:

REMEMBERING EPISODES OF JEALOUSY AND ANGER

↑

MOBILIZING MEMORY, INTROSPECTION
AND OTHER MEANS (

↑

SELF-INQUISITIVENESS
PARTICULAR, INSTRUMENTAL

Both groups of virtues, abilities and character virtues, can be deployed thanks to self-inquisitiveness that looks like fundamental motivating virtue. Here we have to distinguish between instrumental and intrinsic self-inquisitiveness; being inquisitive about oneself, and in particular in one's CD properties, ultimately for the sake of various goods that such self-knowledge can bring to one, and for the sake of self-knowledge itself. The latter, taken in its full scope is an interest in "examined life", in particular in its epistemic form, where self becomes "proper object of knowledge" (Moore 2015: 6).

If you are still skeptical about calling such an interest "-inquisitiveness", you are free to classify it as a separate subspecies of interest, say instrumental interest in oneself (IIIO, for short). Our point then becomes that IIIO is in many cases an important motivating force. If we take seriously the possibility of pathological cases in which even instrumental interest is blocked, we shall understand that such an interest is a virtue, perhaps less 'elevated' than the intrinsic one, but still an epistemic virtue. Then, everything we said can be reformulated as a claim about IIIO.

Let us pass to the intrinsic case. Consider abilities linked to self-knowledge, like self-perceptiveness. Taken in themselves, they don't motivate, similarly to other abilities. I can have sporting abilities, but if I lack a motivating interest, the ability alone will not motivate me. Nor will there be any drive to organize my abilities. I am tall, so I have one quality needed for basketball. Suppose, my motor functioning is relatively precise. Now, the two should come together: I should aim precisely when I am near the target thanks to my tallness. But nothing in the abilities themselves organizes them; only if I have an interest in scoring, will this interest drive them to combine into an efficient whole.

The same thing with self-perceptiveness. It might need to be combined with memory, in order to produce an impressive Proustian saga one one's life. Why would anybody go into combining, if she were not interested in oneself, more particularly in knowledge of oneself? And we end up with self-inquisitiveness.

On the other hand, character virtues do have motivating power. But a character virtue that motivates coming to know about oneself, will come very close to self-inquisitiveness. Take intellectual courage. The kind of courage that motivates one to ask questions about oneself is courageous self-inquisitiveness.

To conclude about intrinsic self-inquisitiveness; being inquisitive about oneself, and in particular in one's CD properties, ultimately for the sake of self-knowledge itself. Taken in its full scope it becomes an interest in "examined life", in particular in its epistemic form, where self becomes "proper object of knowledge" (Moore 2015: 6)

Of course, self-knowledge might be bad in some instances, say de-motivating when it comes to challenges. One reviewer offered the example of over-estimating one's ability at the start of an exam, in order to boost one's morale. But the example nicely shows its limitations. Imagine it generalized: it is good to over-estimate one's ability generally (as a matter of fact i.e. *de re*; one cannot do it on purpose, *de dicto*, since this would involve a pragmatically inconsistent belief, e.g., "My IQ is 80 and I believe it's 130"). Well, such a generalized strategy would lead to disasters; the small advantage of self-assuredness would be vastly negatively compensated by actual disabilities: I can cross the street quickly, so that the car will not hit me, I can jump over this cliff no problem, are just the simplest and most obvious examples of where the (self-)knowledge of one's abilities is precious. I shall leave further discussion of analogous cases aside, and rely on a wide consensus in psychological and philosophical literature that on the whole, self-ignorance is practically a worse option in most typical cases. (But let me point out to Allan Hazlett's 2013 book, as a challenge that I will have to address at some future occasion.)

Back to Othello. For him, the instrumental variety would be certainly enough. Of course, the instrumental self-inquisitiveness allows for a more particularized sub-variety, and the more general one. I can be interested in my particular competences, in order adequately to choose tasks I shall dedicated myself to. I, the present author, love classical music, but I am singularly untalented for singing or playing instruments; once I realize this I should give up any education in these activities, what I indeed did when I was in the high school. Or, I can be interested generally in my abilities, from intelligence to self-control, because I would like to organize my life accordingly. Indeed, a healthy portion of instrumental self-inquisitiveness, both particular and general, would be enough for the needs considered by psychologists, authors like Dunning and Eurich. The latter is focused upon self-insight and self-awareness, "the ability to see oneself clearly" (2017: 13), and argues for the need for it by claiming that it makes us happier in the long run, and that the lack of it might be disastrous for us (ibid.); clearly reasons to investigate oneself for the goods that self-knowledge bring to our life, plus the avoidance of evils that stem from the lack of it. Note that general self-inquisitiveness is in literature often connected to the intrinsic self-inquisitiveness, starting from Plato all the way to Nozick, typically along the lines Renz notes in the passage we quoted above: "to engage in philosophy, they thought, is to explore and *thereby* to ennoble the self" (italics mine).

But why do we need instrumental self-inquisitiveness; is not the goal sufficient to "awaken the abilities and character virtues needed for self-examination", a critic might ask (as the reviewer did; thanks a lot). No, it is not. We can see how it works mobilizing memory and reasoning that lead one to the general conclusion about oneself. In other cases, self-inquisitiveness might be the bridge, needed to pass from the awareness of the goal to the possibly exacting preoccupation with means, and the deployment of abilities needed. "I want to be loved by my teenage daughter; what should I do?" Look at yourself, the way you deploy your authority, your impatience with her little whims, and so on, and so on. Once you become aware of these, you will be able to correct your ways, and your daughter might react in the way you dream about. Psychologists like Eurich, writing manuals of self-help for arriving at self-knowledge use their rhetoric primarily to awaken self-inquisitiveness in their readers' minds.

Of course, philosophers traditionally want more; they are keen on intrinsic self-inquisitiveness, both particular and general, and have promoted the ideal of "examined life", and the insufficiency of the unexamined one. Here is a recent reminder of the Socratic proverb by R. Kamtekar:

> He considers the unexamined life not worth living for a human being. Does examination itself make a life worth living, or is examination necessary for recognizing one's own ignorance, which is a precondition for seeking the ("divine") knowledge of what is fine and good? (2017: 7)

Richard Kraut (2006) also mentions on the one hand the importance of rooting out the errors in the self-image that sounds as an end in itself, and on the other, of arriving at the proper understanding of the good, just and fine:

> The great value of an examined life is that it is the only reliable way to root out errors that have taken hold of one's mind because of defects in one's education, and to fill the enormous gaps in one's conception of what is good, just, and fine. Socrates conceives of it as a process that never comes to an end. (2006: 238)

Of course, the debate about Plato's exact meaning is extremely rich.[12] Here we don't need to go into exegesis; the important point is that some people sometimes are interested in their own CD properties just for the sake of being in the clear about them. For them, the "unexamined life" is not sufficient, and they "test their own soul" to check for properties that interest them. We shall return to the topic in a moment.

Let us summarize the morals of the section: self-inquisitiveness is a motivating epistemic virtue, in fact *the* motivating one when it comes to self-knowledge. It organizes other virtues, and directs them properly.

Consider first the organizing of executive virtues-abilities. I can pass from a piece of perceptual knowledge to a belief, and, by transparency reasoning, to knowing that I have the belief; that would be a relatively simple piece of CD knowledge. But from knowledge about various beliefs that I have and of my manners of acquiring them I can inductively pass to a general conclusion about my habits; say that I am more prone to trust my immediate experiences that the testimony of my closest friends. For this I need inductive abilities, a certain manner of understanding one's habits and reactions, and other intellectual abilities. The general schema might look approximately like as in Fig. 10.1.

We were mentioning specific abilities leading to self-knowledge in the narrow sense. They might include introspection, but when it comes to one's knowledge of one's propositional attitudes I am more in favor of transparency accounts, so I would stress abilities involved in arriving at

[12] Christopher Rowe mentions a reading that is particularly relevant for us here:

> On the interpretation in question (self-examination as the examination of one's belief-sets) this process has to do with examining and sorting one's own individual beliefs, keeping some and throwing others away—a kind of individual intellectual therapy (even if everyone would, presumably, end up with exactly the same, true, set of beliefs).[7] That—on the same interpretation—is what Socrates helps others to achieve, but also, and more importantly, aims to achieve for himself: "Socrates is more concerned with testing his own soul. And he tests it to see if it has true beliefs, assuming that they [sc. beliefs, presumably] determine character" [footnote points to Irwin 1979 (= Gorgias' commentary): p. 182, on Gorgias 486D]. Seen in this way, self-examination is a means of self-improvement, which will—so Socrates hopes—throw up real truths along the way. (2011: 203)

Rowe in fact disagrees with this reading, but the reading is congenial to our questions in the present chapter. Julia Annas, on the contrary, claims that for early Plato, self-knowledge, properly understood, is knowledge of what is impersonal and is most truly real (1985: 136). See also Moore (2015) for a more recent reading.

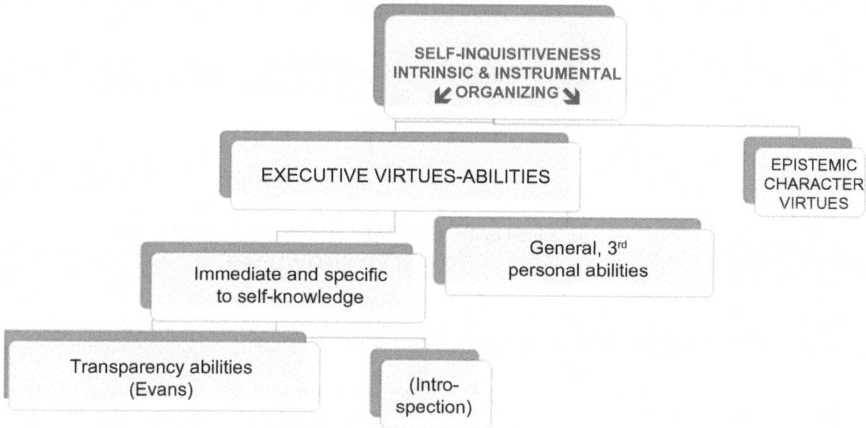

Fig. 10.1 General schema

the relevant pieces of knowledge by asking extroverted question about the world, in the manner of Evans and Fernandez.

In order to lead the thinker to fuller self-attached CD knowledge, self-inquisitiveness also has to mobilize general abilities; our Othello-R should be able to do elementary induction and other forms of reasoning in order to pass from particular memories of his behavior and feeling episodes to a general, approximately true picture of his character. So much for the organizing of executive virtues-abilities.

Self-inquisitiveness, however, also mobilizes the required epistemic character virtues (Fig. 10.2).

This brings us to our next topic, the origin of the intrinsic value of items of self-knowledge.

Self-inquisitiveness and the Value of Self-knowledge

We can be brief about the phenomenal inner-states knowledge. Its practical value is generally clear: Marry sees red (coming from the traffic sign), she is aware that it is red and she stops. But is such phenomenal knowledge of some self-related item σ also intrinsically valuable? Probably, yes. One reason is its constitutive role: no seeing red without knowing that

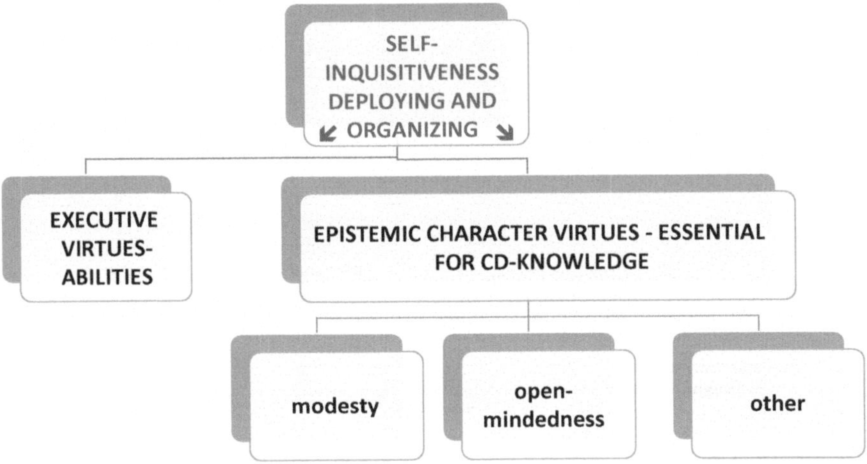

Fig. 10.2 Self-inquisitiveness

one sees it. Another is our natural curiosity. The black-and-white Mary is dying of curiosity about what it is like to see red, others might be similarly curious about experiences in sport or dancing, and, having been reading Lewis in my hometown far away from the United States, I have been intensely curious about what it is like to taste a Vegemite (and got disappointed once I did taste it).

The self-attached CD knowledge has extrinsic instrumental value of all kinds imaginable. On the low, but immensely important level, it enables our survival, and then the fulfilment of our everyday needs and wishes. On the higher level it secures the coherence of our mental "make-up", and gives us a consistent picture of it. Those who are authentic in their virtues, are made us more systematically and intelligently authentic if they enjoy a bird's eye view of their causal powers, active and passive. The traditional ideal of wisdom includes a very high degree of self-knowledge of one's character. Let me briefly mention a fine paper on this kind of value, due to Michael Cholbi. He calls the relevant kind of self-knowledge "substantial", and starts his account with the standard virtue-theoretical line about knowledge as achievement and notes that appeal to achievement explains why self-knowledge has higher value than mere true belief about oneself (web, p. 15 ff.).[13]

Of course, the appeal to achievement also helps explain some of the intrinsic value of self-attached CD knowledge, but it leaves open an important issue: why is specifically knowledge of oneself valuable, and why is it not like pointless knowledge about the number of leaves of grass at some location that is difficult to reach and to inhabit, so that counting grass is a sporting achievement. (Achieving self-knowledge is not just like a sporting victory winning a marathon, to use Bradford's example, in spite of the venerable tradition of comparing achieving knowledge to hitting a target by an arrow). He rightly notes that the distinctive value of self-knowledge must come from it having the object it has. And he introduces the issue of extrinsic, goal directed value under the name of "practical constraint" (web, p. 5). His account is sophisticated, but I don't find it intuitively attractive. He argues that there is a gap between two kinds of loving attitudes we have to ourselves: on the one hand I can love myself for my qualities (he calls it *de dicto* with some additional complications), or I can love myself just as myself, no matter whether I have qualities worthy of love (loving myself *de re*), and that the gap creates a problem. Self-knowledge helps with the gap. It is "valuable because of its role in rationally mediating between the *de re* and *de dicto* aspects of self-love. Our reason to pursue it is to satisfy ourselves that we have good reason to pursue those ends we believe will be sources of happiness for us" (web, p. 28).

One might wonder how relevant this is. If the value is due to lessening of the gap, isn't self-knowledge valueless when the gap remains, or even widens. Suppose I start by loving myself a lot, and also believing that I am a fantastic person, kind, talented and attractive in all sorts of ways. Then I discover that the beliefs are dramatically false, but I still love myself a lot, in fact, as much as before. Then, by Cholbi's proposal, my self-knowledge would be valueless. This is hard to accept: assuming that I am emotionally self-attached and egocentric, it is better that at least I am not narcissistic about my alleged qualities; self-knowledge does a lot of useful work here.

[13] His references are to the book by Gwen Bradford (2015a). But let me note an interesting paper by the same author, Gwen Bradford (2015b).

Let me remind the reader that Cholbi's account is not geared specifically to intrinsic value. In fact, the value that goes beyond the fact of self-knowledge being an achievement seems to be ultimately instrumental. So, we still have to account, in the most elementary case, for the desire one sometimes has simply to find out what one thinks, or deeply feels, and so on, regardless of possible further goals that would be responsible for the additional value of knowledge aimed at. So, we pass to the intrinsic value of the self-attached CD knowledge. A clear symptom that it does have it is that people normally despise systematic self-blindness and complete acedia concerning one's own traits. Interestingly, the attitude is normally stronger against the lack of self-inquisitiveness that against the lack of objectual one. Imagine Jane, a fine person who is volunteering to help refugees. You ask her a question:

> Jane, why did you volunteer?
> God knows, I just felt like doing it, she might answer.
> But why?
> I don't know, I never think why I am doing this or that.

Perfect for a hero from Dostoyevsky or Tolstoy, not perfect for an average university student. We simply expect a degree of self-reflectiveness from people from our surrounding. The self-blind Jane is not like a person who completely lacks interest in politics, in art and classical music and in sports. Such a person is ordinary—Jane is almost pathological. Philosophers-writers of confessions, from St. Augustine to J. J. Rousseau and J. S. Mill, have brought testimony to their self-inquisitiveness, their interest in self-awareness exactly of (what we described as CD kind). It is normal to have a degree of non-practical interest in one's causal-dispositional properties. Self-insight is cherished by people, no matter how difficult it is to achieve, but what would psychologists say? Here is the concluding sentence of Dunning's book on self-insight:

> Life presents many challenges, strewn like hills and mountains in our path. Acquiring self-insight might just be one of those peaks, one that is more rugged and steep than it looks from afar. ...From its peak, one does not know what the view of the psychological terrain might look like, but I

> assume that most people would be quite curious to take a glimpse of this view. (Dunning 2005: 184)

Here is the hypothesis that I would like to propose: the intrinsic value of self-knowledge comes from value of self-inquisitiveness, which is a deep-seated human epistemic virtue. Indeed, the link between curiosity and epistemic value is quite strong, I think. It makes itself felt in our appreciation of all kinds of knowledge. In the case of self-inquisitiveness and self-knowledge there is a connection, but it leaves open the crucial question: where does the link connecting self-inquisitiveness with the intrinsic value of resulting self-knowledge come from? There are two main options, depending on the order of determination:

> First, grasping of some truth about oneself, say knowing, or coming close to know the truth of some CD proposition σ, is intrinsically valuable because a person (with right characteristics) would be curious whether σ. (Response-dependentist account)
>
> Second, persons are justifiably curious whether CD proposition σ holds because reaching the truth about it (say, knowing, or coming close to know) is intrinsically valuable. (Strongly objectivist account)

The usual feeling is that some states of knowing concerning some states of affairs are intrinsically epistemically valuable, and people, if intelligent, well-informed, educated and sensitive, are curious about these states of affairs. (A strong defense of such objectivism about epistemic value can be found in Brady, 2009.)

But why would such a fact have intrinsic value, dictating epistemic axiology? Just postulating that it has one leaves epistemic value unexplained. So, one could choose response-dependent option; self-inquisitiveness bestows intrinsic value on self-knowledge (and a lesser one on true belief about oneself).

Here is one line of thought leading to the response-dependent option. The value of a state of affairs to be grasped is normally experienced as being a property of the very state of affairs and of grasping it. However, value naturalism suggests that it is not an experiencer-independent property of the state of affairs or of grasping. Hopefully, the value is not merely a fiction. Therefore, by principles of charity and by inference to the best explanation, being epistemically valuable in objective sense is being such as to cause the response of intentionally experiencing epistemic intrinsic value in suitable under suitable circumstances. Our graspings of propositions and objectual characterizations are epistemically valuable iff a person, endowed with at least normal cognitive capacities and at least some general knowledge, and familiarity with the domain of p (or, alternatively, the person's somewhat idealized counterpart), would be stably intrinsically curious about p (either whether p is true, or about truths in connection with p, or both). Similar conditions hold for objectual curiosity.

Apply this now to self-knowledge. Here is the general form for accounting for e-value. Assume that σ is a proposition about oneself, in particular about one's CD properties:

> Knowing that σ is epistemically valuable iff a person, sufficiently cognitively normal (or, alternatively, idealized), and familiar with the domain of σ, would be stably intrinsically inquisitive-curious about σ.

Similarly, but with lesser value, for merely believing and justifiably believing that σ.

Now we need a bit of refining. Let me start by introducing a distinction. When Sue learns about herself being too complacent, she experiences the information (and her grasping of it) as being valuable. This is the subjective aspect, and I shall talk of epistemic value-as-experienced, distinguishing it from the objective epistemic value we want to account for (compare it to the experience of surface being colored, in contrast to the objective color). The experience represents the information about oneself (and the grasping of it) as being valuable; the value-as-experienced is being felt as the property of the information (and grasping). I can feel how valuable this thing I have learned is for me, Sue might think. And the experience is transparent; it goes right to the target itself. The epistemic experienced value is transparently present in the target: epistemic experienced value is being experienced as being a property of a state of affairs (a transparency datum). But how do we pass to objective epistemic value? I propose that we do it the same way a response-dependentist about color would do it for colors: a surface is objectively blue iff it presents blue-as-experienced quality to normal observers in normal circumstances.

Analogously, being epistemically valuable in objective sense is being such as to cause the response of noticing epistemic value-as-experienced in *normal/ideal observers* under *normal circumstances* (*response-intentionalism*; for more see Miščević (2004)). Let us now unpack the proposal we just made. As for observers, we have left open two options, the first referring to normal observers, the second to ideal ones. Start with the first, the egalitarian one:

> σ and states of grasping it (justifiedly believing or knowing it) are epistemically intrinsically valuable iff a person H, endowed with at least normal cognitive capacities and at least some general knowledge, and familiarity with the domain of σ, would be stably intrinsically curious about σ (either in whether σ holds, or in truths in connection with σ, or both).

We assume that the cognizer is aware of her cognitive capacities (a small idealization). But there is a catch in the vicinity. People are often inconsistent in their interest and valuations. We obviously have to idealize further; the question is how much. We should probably demand that the person be endowed with (decent to high) cognitive capacities and general knowledge, and familiarity with the domain. Obviously, the proposal needs a lot of work to arrive at the right level of idealization. Too little is unsatisfactory, given human limitations; unfortunately, some people are intrinsically self-inquisitive about worthless properties of oneself. Too much is equally bad: only high level problems will be intrinsically interesting to such epistemically ideal person. In addition, we have the issues of depth and width: short of omniscience, what is the right proportion of going into detail and depth, and wanting to encompass as many areas as possible. So, the general question is with us, concerning both subjects and circumstances: how much idealization and of what kind? I still believe that very probably the intrinsic self-inquisitiveness is the epistemic value bestowing epistemic virtue.

We thus end with a tentative response-dependentist proposal concerning the intrinsic epistemic value of true grasping (belief, knowledge) of facts about oneself: intrinsic self-inquisitiveness is the value-bestowing epistemic virtue. We have tried to address a number of objections to this view, and we hope to have offered at least beginnings of a right response. We concluded that self-inquisitiveness is perhaps the foundational epistemic virtue that bestows intrinsic epistemic value to self-knowledge. The intrinsic value of self-knowledge is thus perhaps response-dependent: since we are self-inquisitive and cherish the trait, we experience self-knowledge as intrinsically valuable.

From Self-inquisitiveness to Wisdom

Let me conclude by pointing again to the connection between curiosity about oneself, our self-inquisitiveness and wisdom. The classical tradition stresses three items of knowledge related to wisdom: know thyself, know particular others, important to you, and try to get to know human nature in general. (An interesting and famous philosophical problem arises here: given these requirements of knowledge, is wisdom compatible with skepticism? We cannot enter it here, just note its relevance.) But why is this important? Well, I should know myself in order to be able to predict how various outcomes will affect me, how I will probably react in various circumstances, in short, I have to get an insight into my causal dispositions, active and passive.

Cicero typically urges that we should strive to know ourselves, "else it will seem that actors have more good sense than us. For they do not choose the best plays, but those that are most suited to themselves .... If an actor, then, will observe this on the stage, will not a wise man observe it in his life?" (Cox, 163). Cicero's "we" is clearly not universal; since he contrasts "us" with actors, "wise man" therefore seems to imply "patrician". But let us return to the role of knowledge in wisdom. It seems to be knowledge of dispositions: the proverbial wisdom is so often couched in explicit or implicit conditionals, prefaced or accompanied by the point of the conditional: He who does a kindness is remembered afterward; when he falls, he finds a support (*Sirach*):

> You should not vouch for someone: that man will have a hold on you. You should not speak improperly; later it will lay a trap for you. (*The instructions of Shuruppag*)

Similarly with knowledge about others: how a particular person is likely to act upon me or upon items I (should) care for, and how in general people act and react? So, *what is primarily needed is knowledge about causal structure in human matters*. In order to be(come) wise I have to know my motives and my habits, my ways of reacting to external events, opportunities and pressures, and about the methods that could change these ways. We have been calling it causal-dispositional knowledge. It has to do with the causal structure of one's self. and one's knowledge about it. One causes things, acting in the world, and also, and crucially important for wisdom and care for the self, acting on oneself. Thus, causal-dispositional-level concerns causally oriented active and passive dispositions of one's self, and thereby the ways of being (possibly) affected by various courses of things, and of reacting to them. (In the philosophical tradition, starting with Plato, an even richer idea of wisdom has been developing: the wise person in this tradition is someone who knows about the universe and the fundament of being. We shall not consider it here; for a brief comment see the concluding chapter, section on philosophical curiosity.)

Keith Lehrer (1997, *Self-Trust*, Chapter 2) has argued that the condition for having a system of judgments solid enough to offer a wise guidance in life is systematically to come up with right preferences (especially the first-level desires). Let us call the reliable source of good preferences by its classical Greek name "phronesis", so that we can talk about phronesis-generated preferences. Now, these phronesis-generated preferences should be action-guiding whenever necessary, since an acratic cannot be a wise person. Call the person having mostly such preferences and thus possessing practical wisdom in the minimal sense, a "phronimos". I would say that this amounts to being a *phronimos* person, performing phronesis-generated actions. The practical component thus encompasses one very important aspect of wisdom, namely moral choice and the role of moral insight or "phronesis" in it.

Indeed, in order to be wise we need to reflect, to have a second order knowledge about our attitudes. Self-inquisitiveness gives us this other crucial piece of knowledge needed for wisdom, the one about causal-dispositional structure of our own character. This is its theoretical pillar helping the practical pillar in holding together the construction.

Here are a few examples from various areas. First, I have to balance my motives and obligations all the time, and for this I might need reflection. For instance, my wish to go out now is balanced by my sense of obligation to finish the paper; as you can see from the thought provoking result you are looking at, the second motive has won the day, and has turned into the resulting preference. But even in this simple case I might have to take a wider view: these are last days with nice weather, and walking and exercise are very important for my health in the longer term. More momentous decisions might demand a wider and holistic balancing that goes beyond the momentary, the local and the molecular, and does happen at the second level. When my university town in Croatia started being shelled in the former Yugoslav war, some of my colleagues and friends decided to leave. They were choosing between staying, being loyal to the colleagues and university but endangering their children (and themselves), and leaving, being in physical security but with a bitter taste of disloyalty, and with an uncertain career perspective. A lot of overarching, second-order elements have been active in their decisions, the view of oneself in a wider perspective, the political view of their town and country in a war and other such global considerations.

Similar global perspective is needed in more traditional moral dilemmas. The problem of "dirty hands" in politics is a good example. An honest political activists knows that the leader of the group is the only one capable of leading to the victory for a good cause; but the leader is otherwise manipulative and corrupt, and his success might lead to the degeneration of relations within the group, to bitter disappointments of the members, and so on. Should she denounce the leader, and risk immediate defeat and catastrophe, or keep silent and risk the long-term degeneration? Well, considerations entering such decision-making might be varied and mediated, demanding a reflective, second order grasp and endorsement, rather than only simple, spontaneous phronesis-guided decision.

Wisdom is a very good trait to have, a virtue for sure. But what kind of virtue? Purely epistemic-intellectual, as Aristotle would have, or straddling the divide between epistemic and practical, as Platonic tradition would have preferred. In the beginning chapters I have divided epistemic virtues into the narrowly cognitive ones on the one side, and the moral-like ones on the other. The narrowly cognitive ones can be divided into motivational and executive ones. The motivational one is inquisitiveness, directed to true and important items, that sets the goal of inquiry: first, getting reliably at truth and avoiding error, second, achieving understanding, that is, grasping important connections. The executive ones are abilities and capacities, perception, intuition, reason, memory and the like. I agree with Greco and Sosa that the executive ones can and perhaps should figure in the definition of knowledge. The moral-like cognitive virtues, like, for instance, openness, intellectual modesty or humility, critical mind and intellectual courage, get part of their value from their cognitive achievements, and part from indicating or promoting a general moral quality: intellectual courage is a sign of courage in general, and so on for modesty and others. They can play both motivating and executive roles (and are being put forward by authors like Zagrebski 1996). I propose to situate wisdom in this hybrid genus; where being hybrid is not meant to be negative, but rather indicative of richness and usefulness. Wisdom has a very strong epistemic side, connected with understanding and knowledge, but also a central practical-executive guiding role.

Let me finish by pointing to the connection with understanding. Reaching understanding is one of our central epistemic goals, dictated by our important motivational epistemic virtue, namely inquisitiveness about the way things hang together. Understanding of humanly important causal dependencies is also the basic factual-theoretic ingredient of wisdom, on the anthropocentric view proposed in the paper. We have considered the more epistemic aspect of wisdom, its intellectual component, setting aside the thorny issues more closely linked with ethics (and moral knowledge, assuming that it exists, an assumption that I would endorse, but cannot defend here). I have argued that most of it concerns causal dependencies and dispositions, and thus actually falls under the umbrella of understanding, especially if we model the relevant kind of self-knowledge in a causal-dispositional manner. Given the epistemic resilience of understanding stressed by authors like Greco, its resistance to luck, and its transcending of mere justified true belief, we can count typical genuine understanding as an instance of knowledge. Wisdom is thus connected with two epistemic goal-states, knowledge and understanding. Taking knowledge to be valuable basically because it secures important truths for us, and taking understanding to consist in knowledge (in strong or weaker sense) of important dependencies, for our purposes most relevantly of causal ones, we have thus arrived at the intersection of the understanding with wisdom.

Conclusion

The main kind of self-knowledge that is difficult to achieve is knowledge of one's causal-dispositional structure, our CD knowledge. It does represent an achievement, in the manner that makes it an interesting target of virtue-epistemological study; virtue epistemology has traditionally been interested in knowledge as achievement. Other kinds of self-knowledge that have been exercising philosophers in connection with topics like physicalism, namely the knowledge of one's phenomenal states, are easy to achieve, in comparison, and we were not dealing with them here.

Epistemic virtues involved in achieving self-attached CD knowledge are many and various, from abilities needed for transparency method(s) to reasoning abilities in general, on the side of virtues-abilities, all the way to the epistemic character virtues. How do these virtues come to cooperate with each other in the acquisition of self-knowledge? We have suggested that self-inquisitiveness is the answer. It is not only *the* motivating epistemic virtue; if successful it deploys, organizes and motivates the use of other epistemic virtues, both virtues-abilities and character virtues. Self-inquisitiveness thus instantiates the general paradigm of curiosity-inquisitiveness that organizes and motivates other epistemic virtues (both virtues-abilities and character virtues).

I have also tentatively proposed that its intrinsic variety is responsible for the intrinsic value of knowledge. Let me put the proposal in a wider context. I have been arguing in Chap. 7 that inquisitiveness or curiosity in general is the motivating and organizing epistemic virtue. It mobilizes and sustains the functioning of both kinds of virtues, abilities and character traits. The goodness of character virtues is two-fold: they are epistemically highly relevant, and they are morally positive, often tightly linked to their purely ethical counterparts. The epistemic side is regulated by the desire for truth, curiosity, and is thus parallel to the epistemic functioning of virtues-abilities. Once we have the general unitary picture, it is easy to look for further similarities: if you believe in fully fledged virtues, that have competence as their integral part, and then we can start comparing these competences to the virtues-abilities, and look for interesting parallels and contrasts. Also, I would be sympathetic to a response-dependentist account of intrinsic epistemic value in general. The account would claim that a piece of knowledge, say knowing that *p*, is intrinsically epistemically valuable iff a person sufficiently cognitively normal (or, alternatively, idealized), and familiar with the domain of p, would be stably intrinsically curious about p. The experienced value is, of course, experienced as being objective, present in the piece of knowledge itself; this is parallel to the case of color and similar response-dependent properties. We see color as being objectively there, on the colored surface—analogously with value. Similarly, our experience represents the information about some piece of interest (and the grasping of it) as being (intrinsically) valuable; the value as experienced is being felt as the property of the information (and grasping). And the experience is transparent; it goes right to the target itself. The experienced intrinsic epistemic value is transparently present in the target.

The same framework can be applied to self-knowledge, and we tried to do it, briefly all too briefly, in the preceding section. It we are right, self-inquisitiveness is a particular, crucially important sub-species of curiosity that enjoys all the powers curiosity in general has: it motivates; it deploys and organizes other epistemic virtues, abilities and character virtues, very probably; and its intrinsic variety bestows intrinsic value on the pieces of self-knowledge achieved.

References

Alicke, Marc D., and David A. Dunning, eds. 2005. *The Self in Social Judgment*. Psychology Press.

Annas, Julia. 1985. Self-knowledge in Early Plato. In *Platonic Investigations, Studies in Philosophy and the History of Philosophy*, vol. 13, ed. Dominic J. O'Meara, 111–138. Catholic University of America.

Baehr, Jason. 2011. *The Inquiring Mind: On Intellectual Virtues and Virtue, Epistemology*. Oxford: Oxford University Press.

Bradford, Gwen. 2015a. Knowledge, Achievement, and Manifestation. *Erkenntnis* 80: 97–116.

———. 2015b. *Achievement*. Oxford University Press.

Brady, Michael S. 2009. Curiosity and the Value of Truth. In *Epistemic Value*, ed. A. Millar Haddock and D. Pritchard, 265–284. Oxford University Press.

Brownstein and Saul eds 2016. Implicit Bias and Philosophy, Volumes 1 and 2, Oxford University Press.

Cassam, Quasim. 2014. *Self-Knowledge for Humans*. Oxford: Oxford University Press.

Coliva, Analisa. 2016. *The Varieties of Self-Knowledge*. Palgrave Macmillan.

Dunning, David. 2005. *Self-Insight: Roadblocks and Detours on the Path to Knowing Thyself*. New York: Psychology Press.

———. 2011. The Dunning–Kruger Effect: On Being Ignorant of One's Own Ignorance. In *Advances in Experimental Social Psychology*, vol. 44. Amsterdam: Elsevier Inc.

Eurich, Tasha. 2017. *Insight: Why We're Not as Self-Aware as We Think, and How Seeing Ourselves Clearly Helps Us Succeed at Work and in Life*. Crown Business.

Evans, Gareth. 1982. *The Varieties of Reference*. Oxford: Oxford University Press.

Ferrari, Michael, and Robert J. Sternberg, eds. 1998. *Self-Awareness, Its Nature and Development*. New York and London: Guilford Press.

Hatzimoysis, Anthony. 2011. Introduction. In *Self-Knowledge*, ed. A. Hatzimoysis. Oxford University Press.

Hazlett, A. 2013. *A Luxury of the Understanding: On the Value of True Belief*. Oxford University Press.

Hoyle, R.H. 2010. Personality and Self-Regulation. In *Handbook of Personality and Self-Regulation*, ed. R.H. Hoyle. Oxford: Blackwell.

Kamtekar, Ramchandra. 2017. Self-Knowledge in Plato. In *Self-Knowledge: A History*, Oxford Philosophical Concepts Series, ed. Ursula Renz, 25–43. Christia Mercer.

Keith, Lehrer. 1997. Self-trust, Clarendon Press, Oxford.

Kraut, Richard. 2006. The Examined Life. In *Blackwell Companion to Socrates*, ed. Sara Ahbel-Rappe and Rachana Kamtekar. Oxford: Blackwell.

Lao Tzu, Tao. 2006. *Te Ching*: A New English Version (Perennial Classics) Paperback, September 5, 2006, translator Stephen Mitchel Harper Perennial Modern Classics; Reprint Edition.

Miščević, Nenad. 2004. Response-Intentionalism about Color [Croatian]. *Journal of Philosophy* 4 (2): 179–191.
———. 2016. Curiosity—The Basic Epistemic Virtue. In *Moral and Intellectual Virtues in Western and Chinese Philosophy, The Turn toward Virtue*, ed. Chienkuo Mi, M. Slote, and E. Sosa, 145–163. Routledge.
———. 2017. The Value of Self-Knowledge. In *Perspectives on the Self*, ed. B. Berčić. Rijeka: University of Rijeka.
Moore, Christopher. 2015. *Socrates and Self-Knowledge*. Cambridge, UK: Cambridge University Press.
Nozick, Robert. 1989. *Examined Life*. Simon and Shuster.
Renz, Ursula. 2017. Introduction. In *Self-Knowledge: A History*, ed. U. Renz. Oxford: Oxford University Press.
Roberts, Robert C., and W. Jay Wood. 2007. *Intellectual Virtues*. Oxford University Press.
Rowe, Christopher. 2011. Self-Examination. In *The Cambridge Companion to Socrates*, ed. Donald R. Morrison, 201–214. Cambridge University Press.
Sosa, Ernest. 2001. For the Love of Truth? In *Virtue Epistemology: Essays on Epistemic Virtue and Responsibility*, ed. Abrol Fairweather and Linda Zagzebski. Oxford University Press.
———. 2007. *A Virtue Epistemology Apt Belief and Reflective Knowledge, Volume I*. Clarendon Press.
———. 2015. *Judgment and Agency*. Oxford University Press.
Stalnaker, Robert C. 2008. *Our Knowledge of the Internal World*. Oxford: Clarendon Press.
Ulrich Wiesmann, et al. 2013. Validating the Grand Five of the Generalized Health-Related Self-Concept. In *Psychology of Self-Concept 2012*, ed. Kamel Gana, 7–20. New York: Nova Science Publishers.
Zagrebski, 1996. *Virtues of the Mind*. Cambridge University Press.

11

Conclusion and Tasks Ahead

What We Did in the Book

The Central Chapters

In this chapter we shall do two things. First, we summarize the main claims of the book, and then we briefly sketch a few directions in which philosophical discussion of curiosity might and, in our view, should develop. The summary will be done following the content of the book, chapter by chapter, and the directions for the future development will differ to a great extent in their content.

Let us start with a bit of the background. Virtue is again the topic of the day, both in ethics and in epistemology. Virtue-epistemology has been looking up to virtue ethics for guidance, partial or total. However, there are important sources of tension within the virtue epistemological paradigm. Epistemology is traditionally concerned with knowledge and stresses the importance of truth, of one's reliable arriving at truth and being able to tell oneself whether one is thus reliable. Truth is central, whereas the positive qualities, "virtues", of the cognizer are just means to arrive at it. The view that takes seriously this traditional and quite

N. Miščević, *Curiosity as an Epistemic Virtue*, Palgrave Innovations in Philosophy,
https://doi.org/10.1007/978-3-030-57103-0_11

plausible view of the enterprise of cognition takes virtues as being important but not as being fundamental. Ernst Sosa is the pioneer of this approach (see Sosa 2001, 2007). Naturally, the virtues involved are more like executive qualities than like traditional moral virtues that primarily motivate the agent to pursue her goals, and only secondarily aid it in pursuing it. Furthermore, the central role of success in reaching truth stands in stark contrast to the relatively modest role of success in traditional virtue ethics, which concentrates more on the right motivation and intention plus honest trying than on external success.

On the other hand, a serious involvement with virtues, in particular moral ones, would prompt, and has prompted philosophers to look for motivating, moral or moral-like virtues in the domain of cognition. They might even propose that there are "ethical foundations of knowledge", as the subtitle of Zagzebski's 1996 book *Virtues of the Mind* suggests. But epistemology does not seem to be just a branch of ethics. So it seems that truth and virtue don't go well together in epistemology, and that if you maximize the value of truth, you minimize the one of virtue, and the other way around. Moreover, even most of the other proposals within this camp, except Zagzebski's one, stop short of claiming that the values pursued in epistemic enterprise are themselves dependent on virtues. Rather, virtues enable their bearers to realize or implement the values independent from virtues themselves. Indeed, how could virtues ground or produce the value of truth, or understanding? So, even in this 'moralizing' camp one rarely finds a virtue-based approach. Most philosophers in both camps are thus content with a mere virtue-focused line, and a virtue-based project is a bit too much for them.

We have been trying to show here that we can do better, that we can have a "strong" virtue-epistemology, to borrow Blackburn's term for epistemology that is based upon virtue, instead of only taking interest in it. The main tension within virtue epistemology can be overcome, and that it makes sense to try to work on a version that would be at the same time virtue-based and truth-centered, thus first taking the best from both camps, and then building up the role of virtue in the resulting picture to its full strength, sufficient to ground the epistemological project. The secret of obtaining a virtue-based account lies first in picking up the only motivating trait that has to do with truth, namely inquisitiveness or curiosity, and second, in promoting it to the status of the central motivating epistemic virtue. Truth and virtue after all, should go well together in epistemology. On the other hand, virtues-capacities discussed by authors like Sosa and Greco are executive virtues that help reach the goal set by inquisitiveness, while other epistemic character-trait virtues, like open-mindedness or intellectual honesty, require a separate treatment. The third move is then the obvious one: indeed, we have argued in Chap. 7 that truth has value because there are cognizers interested in reaching it. If this holds, the virtue of inquisitiveness is the one that bestows value upon truth, and we have a virtue-based framework for speaking about knowledge. None of this need enter the strict definition of knowledge, nor would it tend to 'moralize' epistemology. Sosa's paradigm of defining knowledge by appeal to executive virtues seems to be exactly what is needed in the narrow conceptual analysis of knowledge. The proper place of appeal to the motivating virtue(s) is in accounting for our desire to know, and in analyzing the goal of epistemic enterprise, not within strictly defining its product. Since the view proposed integrates motivating and executive epistemic virtues, it is an integrated virtue-based view; I hope it deserves its name.

In Part I we started with briefly retelling the historical origins of discussions of curiosity. It has turned out, somewhat surprisingly, that the pioneering authors, from Socrates and Plato to late Hellenistic philosophers, have paid a lot of attention to distinguishing kinds of curiosity, and offering different evaluations of each of the various kinds. Our topical task, of taxonomizing kinds of curiosity, has turned out to be a direct continuation of this rich historical effort!

We then first look at the central epistemic role of curiosity, and then to its function in human psychology, and in search for happiness and meaning. Along the way, we stress its connection to understanding, and also sketch its relevance to moral issues and meaningfulness of life, focusing on curiosity about oneself, or self-inquisitiveness as the typical sub-species incarnating a special relevance for human life.

The main contrast relevant for the central part of the book, is the one for and against curiosity. People like Samuel Johnson wrote aphorisms in favor of the first stance, for curiosity:

> Curiosity is, in great and generous minds, the first passion and the last.–
> Curiosity is one of the permanent and certain characteristics of a vigorous mind. (The Rambler, Aug. 24, 1751)

Others, equally famous, took a stance against it:

> Ambition and curiosity are the two scourges of the soul: the latter prompts us to poke our noses into everything; the former prevents our leaving anything in doubt or undecided.—Montaigne. (Essays 1580)

We have decided for the first stance. So, at the beginning of Part II, we address the strongest traditional argument against the positive status of curiosity, namely the existence of cases of bad curiosity. We point out that this phenomenon is valid for many virtues, and analyze the case of bad courage as the central example. We conclude that the strongest argument is not strong enough. We then propose a master argument in favor of curiosity being not just a virtue, but the central cognitive virtue:

1. It is epistemically very important to be intrinsically motivated to acquire knowledge and understanding. Then, we note that
2. most of the character virtues apart from curiosity do not motivate such acquisition, but rather control, in epistemically and morally relevant ways, the process of acquisition undertaken.
3. Virtues-abilities are not motivating in themselves; they help realize the goals we are independently motivated to achieve. And we know that
4. the genuine curiosity is the central intrinsically motivating drive for achieving knowledge and understanding.
5. So, if curiosity is an epistemic virtue, it is *the central cognitively motivating virtue.*
6. Now, it seems that a central motivating virtue, if available, would be basic for the whole system of epistemic virtues (and their contraries).

If this seeming is correct, we can appeal to the fact that curiosity is an epistemic virtue, and conclude that

7. curiosity is the central epistemic virtue, the cornerstone of the whole cognitive enterprise.

Let me point out how Baehr-Zagrebski's characterization of character virtue nicely fits the structure of our Master Argument. Jason Baehr in his 2011 book proposes that a virtue "can be defined as a deep and enduring acquired excellence of a person, involving a characteristic motivation to produce a certain desired end, and reliable success in bringing about that end" (137). He refers to Zagzebski and states:

> She says that a "motivation" is "a disposition to have a certain motive" and that a "motive" is "an emotion that initiates and directs action to produce an end with certain desired features". (136)

And then he continues: "as this initial characterization suggests, Zagzebski anchors her account of virtue in the concept of a good or virtuous motive; thus it is aptly labeled a 'motivational' account of virtue" (2011: 133).

Consider now our Master Argument. It is organized around the motivating force of curiosity, and the Baehr-Zagrebski's characterization, to be applied primarily to epistemic virtues, similarly focuses upon the motivational role of virtue. They nicely supplement each other.

A lot of counter-arguments has been proposed against the epistemological relevance and general value of curiosity, by prominent authors like Sosa and Brady The central paper is Sosa (2001). We address these arguments in a separate Chap. 3 with Sosa as the main interlocutor.

The chapter on epistemic value (Chap. 7) addresses two fundamental issues in epistemic axiology. It argues primarily that curiosity, in particular its intrinsic variety, is the foundational epistemic virtue since it is the value-bestowing epistemic virtue. A response-dependentist framework is proposed, according to which a cognitive state is epistemically valuable if a normally or ideally curious or inquisitive cognizer would be motivated to reach it. Curiosity is the foundational epistemic virtue, since it bestows epistemic value. It also motivates and organizes other epistemic virtues, so it is foundational and central for epistemology.

The second issue is the one of the fundamental bearer of epistemic value. I argue that truth is the primary goal, but that mere true belief is not the fundamental bearer. Rather, the bearer is a relatively minimalist kind of knowledge. Mere true belief cannot be rationally accepted in isolation from a supporting structure. However, any efficient supporting structure introduces further epistemic goods (justification, reliability, anti-luck guarantees), thus upgrading the original true belief. Mere true belief can be neither defended, nor rationally sustained through time, due to the isolation. Mere true belief cannot be rationally sustained in the face of a slightest bit of contrary evidence (the Meno insight). Therefore, mere true belief is not rationally stable. In contrast, minimal knowledge is stable, and this accounts for primary and secondary value problem, and for a relatively undemanding kind of tertiary value.

In the last chapter of Part II we go a step further, and propose that intrinsic curiosity is responsible for the intrinsic value of knowledge and understanding: the latter are response-dependent on the former. This somewhat speculative proposal concludes the central part of the book, and the presentation of our theory of curiosity as the central motivating epistemic virtue.

Applications and Widenings

The Role of Understanding

Part III brings together applications and widenings of our approach, centered on three topics. First, the topic of cognitive psychology of curiosity, next of the role of curiosity in science, and finally of curiosity about oneself, or self-investigativeness.

One interesting phenomenon linked to the widenings, is the fact that in all three domains it is the desire to understanding, or understanding-directed curiosity that plays the main role; the analogous phenomenon will appear when we turn to further topics, in the second part of the Conclusion, to further topics awaiting an analysis. So, let me re-iterate what I said about understanding in the text.

There is an agreement in the recent literature on understanding that it has to do with seeing connections, in particular the causal ones. And this seeing has two faces: on the one hand, it responds to the theoretical interest why, or how these phenomena hang together. On the other, it often satisfies some practical interest: in many cases, once you know why, acting on right variables, you can influence the process, thus influencing nature or influencing other people. This is also central for the social epistemology of inquisitiveness; the social organization of research is often application-directed.

In my view, understanding in general is a species of knowledge—I agree with Grimm (2011), as against Pritchard (2010), Elgin (1996: 206), Zagzebski (1996: 201), Kvanvig (2003) and Wayne Riggs (2009). Its epistemic value derives from its richness, cognitive relevance and role in manipulating causes.

Let me remind you of the work of Jonathan Kvanvig, the only author who has explicitly raised the question of understanding-driven curiosity: moreover, he sees it as central to his theory. Here is his list of values tied to understanding-directed curiosity:

> [T]o have mastered such explanatory relationships is valuable not only because it involves the finding of new truths but also because finding such relationships organizes and systematizes our thinking on a subject matter. (2003: 202)

> Such organization is pragmatically useful because it allows us to reason from one bit of information to other related information that is useful as a basis for action, where unorganized thinking provides no such basis for inference. Moreover, such organized elements of thought provide intrinsically satisfying closure to the process of inquiry, yielding a sense or feeling of completeness to our grasp of a particular subject matter. (Ibid.)

He stresses the "objectual" understanding, not of isolated statements but of a larger domain as a whole (169), the typical experience involves objectual understanding of some topic or issue or phenomenon (170):

> [O]bjectual understanding is the central goal of the phenomenon of curiosity. (2012: 170)

I do not share this view: speaking about Inan, I have stressed that objectual curiosity very often involves propositional one, and there is no reason to give primacy to the objectual in relation to the propositional.

So, let me conclude with a few more words about my view of understanding-directed curiosity. Two lines are relevant. First, the actual importance of what is found-understood: Epistemic value of understanding derives from its richness, its organized and systematic character (Kvanvig), cognitive relevance and connecting to distant consequences (possibly related to action, as Kvanvig notes). But the understanding-focused curiosity might also follow, or at least be enhanced by deep-seated preference for and interest in manageable complexity. Psychologists, (Paul Silvia is the best example) stress that people are repelled by too simple, and too complex cognitive task: they become interested in complex but still manageable proposals. So much about understanding as the goal of curiosity.

Let me now pass to the particular chapters, from the eight to the tenth. Psychologists (from Berlyne to Silvia) suggest that it is the novelty of the problem (or of the item to be investigated), and its complexity (hopefully manageable), that selects problems individuals choose to address. We have taken Silvia as our cognitive-psychological guide to the nature of curiosity. He investigates why people learn and explore in the absence of obvious external rewards and talks of curiosity as of "a motive to reduce negative states, such as uncertainty, novelty, arousal, drive, or information gaps" (2012: 157), and stresses its role of a "source of intrinsic motivation that fosters learning and exploring for their own sakes" (ibid.). This fits well with our motivating account to the effect that curiosity is mobilizing and organizing, or helping to organize epistemic virtues, both character-virtues and abilities.

We have translated Silvia's idea that it is "coping" with phenomena that provokes to a large extent people's curiosity, by interpreting "coping" in central cases as understanding.

Read is such a way, the psychological research offers a view of curiosity as being primarily a desire to understand, or, for less mature cognizers, as a desire for some kind of elementary proto-understanding. This offers a fine opportunity for naturalizing the desire to understand, the most sophisticated kind of curiosity. On one level, it offers a scientific grounding for the talk of such a desire, on the other, it points to goals and function of the desire and its typical object, thus helping to see what makes the normative-teleological point of such activity (see also the section on naturalization, Sect. 2).

Science

The chapter on curiosity in science argues that it is curiosity that mobilizes, steers and to significant extent organizes scientific efforts—a fact almost never really analyzed in the literature, nor used in understanding the history of scientific research.[1] The curiosity in scientific practice is the motivating feature, engaged in deploying, focusing and helping organize our knowledge-capacities. In line with this conjecture, I point to varieties of curiosity, noting how different varieties drive the scientist to different domains and different activities.

The topic of scientific curiosity covers a lot; its domain is coextensive with a large part of philosophy and methodology of science. A number of books deals with understanding in science, and they are relevant for our interest in curiosity, in particular the understanding-directed one. Others deal with division of scientific labor, where the distribution of investigative tasks is one central topic. Then, there is a huge literature on history of science, in particular on scientific revolutions, which on scattered places addresses the question of the kind of interests responsible for revolutions.

Scientists themselves have come to understand this central role of their curiosity. As Einstein noted:

> I have no special talents. I am only passionately curious. To Carl Seelig, March 11, I1952. (Einstein Archives 39-013, Albert Einstein (2010). *The Ultimate Quotable Einstein*, p. 20, Princeton University Press)

[1] The new 2019 volume by Pennock does point in this direction, however.

But why is curiosity so important for you, Professor Einstein? Here is what I would say if I were in his shoes: In our ordinary life it organizes our cognitive capacities, deploys them and turns them to particular issues and topics. And it does the same in scientific practice: it is the motivating feature, engaged in deploying, focusing and helping organize our knowledge-capacities.

As a consequence, the study of curiosity-investigativeness and its kinds can help understand the cognitive structure of scientific inquiry (theoretical, applied, hybrid, "high", "low", etc.), it can help systematize historical research concerning science (for example, and prominently, the cognitive structure of scientific revolution with individual scientists and "societies"). It is typically not addressed in this way. For instance, we noted in the chapter on curiosity in science (Chap. 9) that the two extant studies with "curiosity" in their title, the one by Ball and the other by Huff don't go beyond collected facts. It can help systematize the rich relations between theoretical and applied science, and their motivational underpinnings.

One outstanding task, not yet properly addressed in the literature, is to explain the role of scientific understanding, Darwin's *"keen pleasure in understanding any complex subject or thing"*, from traditional to contemporary, the kinds of motivation-to-understand, and the way it has been guided research in some area or other. We would assume that epistemic value of scientific understanding derives from its richness, its organized and systematic character. the latter enhances its cognitive relevance and enables the scientist to connect local discovery to its distant and up to the present moment unexplained consequences. Finally, it plays the central role in manipulating causes, thus steering technological application. Again, in the central case of the search for understanding, it is curiosity that most often mobilizes, steers and to significant extent organizes scientific efforts

Self-Inquisitiveness and Wisdom

We quoted Socrates and his claim that "unexamined life is not worth living" (*Apology*, 38a). The examined life is a life of knowing oneself, recommended by the oracle, which seems to recommend to us a life of permanent self-inquiry, and attitude of permanent self-inquisitiveness. A much more radical, but in some respects limited line in favor of the intrinsic value of knowing oneself is present in the Aristotelian tradition, with the ideal of the Intellect thinking (about) oneself; it is not a kind of self-knowledge we would think of today, but it is worth mentioning.

In Chap. 10 we concentrated upon knowledge of one's causal and dispositional properties ("CD properties", for short), like my knowledge that I am a gourmet, or that I am prone to jealousy. The choice is easy to justify: this kind of self-knowledge has been the focus of traditional and tradition-inspired reflection on self-knowledge, from Greeks to Foucault, and is again prominent in the work of the authors like Qassim Cassam.

For humans, it is often the case that a complex situation is affecting one. Here, causal connections to oneself are much more complicated than in elementary case, but it is again causal structure that counts and the agent needs a reliable model of these causal structures, with self-knowledge as an important focus. Again, one is reacting on the basis of expectations of causally organized course(s) of events, this time starting from oneself; time for contingency planning, reacting in thought to imagined, possible situations. The deeper the agent goes in self-locating and self-concerning thought, the richer and more interesting the causal structure gets.

As before, the action end of immediate self-knowledge (awareness of intention, and then also of desire) is crucially connected to experiencing oneself as a (potential) cause. Causal connections are much more complicated than in elementary case, but it is again causal structure that counts. Here I, the agent, need a modally rich and flexible view of myself, which is exactly what developed self-(CD-)knowledge is supposed to offer. And again, I am reacting on the basis of expectations of causally organized course(s) of events affecting me or starting from me. Central to the CD level is the causal structure of one's self. One causes things, acting in the world, and also, and crucially important for wisdom and care for the self, acting on oneself. So, the CD level concerns causally oriented active and passive dispositions of one's self; the ways of being (possibly) affected by various courses of things, and of reacting to them. Agents have at least three kinds of reasons to look at self-(CD-)knowledge: first, a wide range of practical applications, from survival to small needs and pleasures.

Second, the practical importance of "self-critical perspective", crucial for ethics and views of human happiness and welfare. The point of it all might be the care of the self, finding and realizing the most meaningful kind of life for myself. Self-(CD-)knowledge obviously has a wide range of practical application from survival to wisdom. (On the theoretical side we have at least two kinds of motivation. First, the traditional philosophical interest: making sense of classical views of self-knowledge. Second, the interest of psychology as science.)

Note that in normal case one's self-knowledge interacts with one's knowledge of the social surroundings, since the causal chains very often extend in this direction. The practical value of such a wider knowledge is crucial for our practical interests.

Our motivating virtue account joins the Socratic tradition. It claims that inquisitiveness or curiosity is the motivating epistemic virtue. In the case of self-knowledge, self-inquisitiveness intrinsic and instrumental, is the motivating epistemic virtue that mobilizes other virtues, skills and epistemic character virtues, needed to achieve such knowledge. Its proper object is substantial self-knowledge, knowledge of one's dispositions and causal powers that has historically played central role in philosophy, and is now, under various names, investigated by psychologists. It has been, until recently, comparatively neglected within analytical epistemology of self-knowledge. Self-inquisitiveness thus instantiates the general paradigm of curiosity-inquisitiveness that organizes and motivates other epistemic virtues, (virtues-abilities and character virtues.) And it is perhaps responsible for intrinsic value of self-knowledge.

The Tasks Ahead

The proposal will be again centered on understanding, and exploring the role of the desire to understand (or understanding-directed curiosity) in important domains of cognition, and of human life in general.

Philosophical Curiosity

We wrote about scientific curiosity. We also wrote about the Socratic self-inquisitiveness. In both cases the connection to philosophical curiosity is obvious. First, the deeper questions with which science sometimes deals are also topics for philosophy. Second, the Socratic self-inquisitiveness is a fundamental part of the program of the philosophical desire to know and understand.

So, a natural direction in which to go next is to turn to philosophical curiosity. What is typical of this kind of curiosity? As one would expect it is understanding directed. And it normally involves philosophical questioning as a life project; it is a long-term curiosity engagement. The questioning has an interesting social dimension. Philosophy has been born in the the dialogical tradition, and has lived through various kinds of quasi-institutionalized debating. On the contemporary scene there is the hermeneutic tradition stressing the dialogue, and the analytic tradition of adversarial mode dialogue, with mutual criticism as a central tool.

But philosophy is also specific in one fundamental respect, namely in its tendency to link intrinsic, theoretical curiosity to high extrinsic goals like finding the meaning of life, the nature of the morally good life, and prudentially perfect life.[2] The interest, the desire to understand, is invariant across history, one finds it in Socrates, early modern skeptics, Hume, Kant, all the way to our time, and in our time, common to continental philosophy (with Heidegger, hermeneutical tradition, Sartre and Camus) and analytic one (from Nozick to Duncan Pritchard). A philosopher typically tries to grasp essential connections at the general, categorical level bringing together two levels normally very much disjoint in science and in everyday life: categorial (-scientific, metaphysical) and personal, existential. Their disjointness is dramatic: the latter is hardly noticeable in science, the former hardly present in everyday life.

[2] I owe this understanding of philosophical curiosity to some extent to my friend and colleague Danilo Šuster, who has published in 2019 a fine book (in Slovenian) on "what philosophers are doing". This section owes a lot to my conversations with him. He noted a phenomenon we all encounter in teaching: good philosophers, from Socrates on, try to connect the understanding of personal, existential matters with the understanding of most general categorial (-scientific, metaph.) matters, and to understand the connections themselves.

Notice how Plato and the character of Socrates in his dialogues achieves the grasping. Take any dialogue, for instance *Euthyphro*. It begins with a practical moral issue: Euthyphro has been prosecuting his father for murder of a slave. Now, people say that it is morally wrong, in this context, that "it is unholy (ἀνόσιον) for a son to prosecute his father for murder" (4e). Who is right, Euthyphro or his critics?

Socrates passes from this concrete question to a more abstract consideration: what is common to all holy acts, and to all unholy ones:

> 6d Now call to mind that this is not what I asked you, to tell me one or two of the many holy acts, but to tell the essential aspect, by which all holy acts are holy; ...

Very soon in conversation the question will arise of what makes something holy:

> 10e [W]e are agreed that the holy is loved because it is holy and that it is not holy because it is loved; are we not?[3]

With this turn we are deep in the most abstract issue of whether moral-religious properties are dependent on our responses.

Similar turn is found at the beginning of the *Republic*, when the most general issue of the nature of justice is introduced a propos a commonsensical remark of the old Cephalos to the effect that "the man who finds many unjust deeds in his life often even wakes from his sleep in a fright as children do, and lives in anticipation of evil" (331a). He continues contrasting such a person to a just person: "To the man who is conscious in himself of no unjust deed, sweet and good hope is ever beside him-a nurse of his old age, as Pindar puts it" (ibid.).

Socrates immediately asks for a definition of justice, and in a few sentences the discussion turns to abstract matters of the nature of justice. Later parts of the dialogue go much further. We noted the centrality of self-inquisitiveness for wisdom in Socrates' original project. In the

[3] Plato. Plato in Twelve Volumes, Vol. 1, translated by Harold North Fowler; Introduction by W. R. M. Lamb. Cambridge, MA, Harvard University Press; London, William Heinemann Ltd. 1966.

philosophical tradition, starting with Plato, an even more rich idea of wisdom has been developing: the wise person in this tradition is someone who knows about the universe and the foundation of being.

Indeed, in both Plato and Aristotle the attunement to cosmos, or even deeper, to its metaphysical foundation, secures the crucial goods for human beings: meaningful life, prudential, moral and political goods. This Greek philosophical desire for truth (positive curiosity) directly connects with interest in the highest grade knowledge of value matters we listed. Understanding is the main goal, with depth, and connectedness in breadth as desiderata. The goal involves high levels on quality matters: active character, intensity, lifelong desire, and strong connectedness with epistemic character virtues (and other virtues in general).

Let me briefly sketch my own view of the further development: take it as a brief proposal, since this is all that can be offered in a limited space.

The new problems that arise in modernity preserve the ambition to understand the personal and existential together with the deepest foundations of reality. While science is revealing a humanly neutral cosmos, philosophers try to regain the ancient unity of understanding. Think of the role of god in Descartes and Leibniz, as well as nature-god in Spinoza. Modern philosophy looks at the ways of intertwining of theoretical and practical. German idealism drives speculation back to unity, with special place for man and for our history (Hegel, Nietzsche and Heidegger) bringing two levels together: categorial-metaphysical and personal, existential.

Let me illustrate with an example from Heidegger. It was Heidegger who turned phenomenological investigation into the analysis of the existential relation between Dasein and Sein, and then into a poetic-hermeneutic investigation into human destiny. He started in *Being and Time* with the idea of human involvement with the world, as an antidote to skepticism. (One route from there is the pragmatist one, taken by many of his American interpreters.) In later works the involvement is characterized as "living poetically" (dichterisch). Our motto "Being's poem, just begun, is man", taken from his *Aus der Erfahrung des Denkens*, combines all the elements we were talking about. First, the idea that human being belongs to the very ground of being, that it is ontologically most intimately connected to it. Second, that the relationship between the two is primarily poetic, as opposed to say, epistemic, or logic. Man is the poem, *Gedicht* of Sein. Which reminds us of the idea that "poetically dwells the man on the Earth", taken from Hölderlin, and philosophically developed by our philosopher. And of course, the philosopher is expressing this in a poetic way, not in cold theory, nor in a sequence of arguments.

Just in case one might think it is an isolated fragment, let me give its context:

> When the early morning light quietly grows above the mountains .../
> The world's darkening never reaches to the light of Being./
> We are too late for the gods and too early for Being. Being's poem, just begun, is man./
> To head toward a star—this only./To think is to confine yourself to a single thought that one day stands still like a star in the world's sky. Heidegger (1971: 4)

Let me add a few words on continental philosophy, just to avoid misunderstanding. Indeed, we should distinguish two trails of continental philosophy. One trail is argumentative, and strongly argumentative indeed. This is the trail that has developed out with neo-Kantianism and with Brentano and his followers, from Meinong to Husserl in the second half of the nineteenth century. These were great and important thinkers but they are often seen as atypical for the mainstream continental philosophy. Let me therefore call it low-profile continental philosophy. Let me mention the neo-Kantians, Meinong, Bradly, Husserl, early phenomenologists, and Dilthey and his followers. It is characterized by a positive interest in science, cold, flat and argumentative style, close to the one of their analytic colleagues. So, at the turn of the nineteenth/twentieth century, from 1880s to the 1920s, there was no dramatic contrast between the fledgling analytic philosophy and its continental counterpart. There was exchange and interaction: Frege-Husserl, Russell-Bradley and so on and no dramatic contrast. And there was little influence of the flamboyant, poetic writers like Kierkegaard or Nietzsche in German-speaking academic philosophy. No wonder, this low profile continental philosophy is nowadays close to the heart of the 'bridge builders", from Mulligan to Zahavi. Of course, some later continental philosophers are argumentative and practically come close to the low profile, the most famous ones being early Sartre and Merleau-Ponty, Gadamer, Ricoeur and Habermas.

On the other side, there is the opposite tradition within continental philosophy. It has some of its roots in Hegel; not that he hated argument, not at all, but his view of logic was very special, and didn't encourage a development of argumentative philosophy. But the real, typical nineteenth-century authors in this trail are Kierkegaard and Nietzcshe with their literary ("*Either-Or*") or poetic ("*Zarathustra*") writing. In the twentieth century Heidegger made it enter the academic scene in Germany, and made it the central trail in the philosophical academia. Today, this tradition is seen as what is typically coninental: Heidegger, Adorno, Benjamin, later Merleau Ponty, Derrida, Deleuze, Žižek. Let me therefore call it the "high-profile" continental philosophy. Here we encounter the typical, clear and dramatic contrast with analitic philosophy, often taken as central in antagonistic contexts.

Let us return to our general short sketch of philosophical curiosity. If the sketch is even approximately valid, a large part of philosophical curiosity as we know it from history of philosophy is a special, highly structured variety of the desire to understand. The analytic philosophy is more cautious in this respect. It preserves the unity of method but keeps disunity of content (target) with theoretical issues separate from practical ones.

I am aware that the sketch is very short, and that the topic would require a chapter, if not a book, but we have to be brief.

Social Epistemology of Curiosity

The second direction we want to point to is the direction of social knowledge, and social organization of curiosity-investigativeness (which we briefly mentioned in Chap. 3). We live in a world of significant epistemic division of labor: specialists versus non-specialists, theoreticians versus practicians, institutional accusers versus institutional defenders in law and in media, information-seekers versus information-providers, and so on. The epistemic work is central to our society, and in particular to its economic and political functioning. The division of epistemic labor increases efficiency, and creates now possibilities, but it also adds to existing inequalities, and creates new ones, from economical, through legal to political ones.[4] When it comes to social organization of investigation, virtuous curiosity should cater for epistemic justice; the topic of epistemic justice is therefore a burning topic. Its core are issues of fairness and equality, and it has highest relevance for public reason.[5] Individual curiosity has been the paradigm for mainstream epistemology, both classical and contemporary; in particular, the virtue(s) discussed belong typically to the varieties of individual curiosity. Social curiosity offers new topics. First, there is a wealth of relevant areas: science, law, media, education, politics.

Thus objects vary, from low thought medium to high ones (sex scandals vs. everyday politics vs. cosmological theories). Education can help raise the level of object. Also, in contrast to Antiquity in modernity it is normal to have high knowledge of low objects (science investigating viruses and bacteria, courts professionally looking deeply into crime and sex). Similarly with the levels of understanding: in media communication it is often, minimally required: reporting an important event as it happens, without being able to figure out causes and consequences (only requirement is the amount of undrstanding needed to identify the object). Only at higher level, for instance, when it comes to professional comments, is it central.

The bearers equally very. And individual might be the bearer, of a social item, group and the like. The social bearer is either non-institutionalized or institutionalized; and the contrast is crucial for understanding of many social epistemic phenomena. Here is an example of demands put by society to the individual bearer:

> An investigative reporter needs to have
> Curiosity
> Passion

[4] We already mentioned the crucial and pioneering work in this area, Goldman's 1999 book *Knowledge in the Social World*, Oxford University Press.

[5] See Chambers (2009), "Who shall judge? Hobbes, Locke, and Kant on the construction of public reason", *Ethics & Global Politics*, Vol. 2, No. 4, pp. 349–368.

> Initiative
> Logical thinking, organisation and self-discipline
> Flexibility
> Good teamworking and communication skills
> Well-developed reporting skills
> Broad general knowledge and good research skills. (Konrad Adenauer Stiftung: *The Investigative Journalism Manual* (2016: 11 ff.))

Similarly, the driving force might take the bottom-up direction, mainly in investigator-driven inquiries, or go top-down, involving the state or some smaller institutions.

Motivation goes from extrinsic to intrinsic, and from pure to applied research. Specialization in intrinsic motivation sometimes supports pure research. However, it can be quickly coordinated with application and embedded in applied framework. The most famous and most dramatic historical example concerns early nuclear physics. The idea of the bomb became crucial because of fear from Nazis; soon we get nuclear research financed by the army, with events like Hiroshima and long-term processes like the Cold War.

Values in play can be epistemic, moral, or prudential: egoistic/group altruistic/general altruistic, value of meaningfulness, of political (justice), legal value and overall value. They yield normative and value status of social inquisitiveness. Consider the contrast of meaningful, problematic and meaningless, and its application in social epistemology of media. Noel Carrol write in *Media Ethics* (M. Kieran (ed), Routledge, 1998):

> If we watch too much television that is probably a function of the fact that in our culture we do not spend much time training people how to use the medium: how to integrate it in a fulfilling life-plan. This, of course, may itself reflect an unwillingness in certain societies to include as part of basic education thinking about how one might lead one's life and the habits one needs to develop to pursue such lives. The problem of excessive television viewing, then, is a moral problem, if it is a moral problem, because of a larger cultural failing. It is not an ailment to be attributed to the medium as such, nor to its characteristic stylistic elaboration of the image. Neither the medium nor the image is inherently immoral, though our systematic failure to educate people about how to use it may be socially irresponsible. (p. 151)

Two more examples of negative evaluation are worth mentioning. And we did, indeed, already mentioned that Foucault made famous Bentham's panopticon as a spectacular example of bad extrinsic value. The other example is the mainstream continental evaluation of science; from Heidegger to Habermas it is seen as an endangering phenomenon.

Next, there is the central role of cooperation and competition: typically, a relevant topic becomes distributed between researchers (some do more theoretical aspects, other more experimental ones, and the third help with technicalities), the attention is thereby enhanced, and then the results are balanced and brought together.

Distinguish between non-cooperative and cooperative scenarios. Within the first, we can further distinguish vigilance as attitude and competition as the fundamental structure. Vigilance prompts checking and often rejecting cognitive proposals from others. Competition leads to stratification/and very often to specialization. On the cooperative side we can also have specialization, combined with coordination. In the less advanced case, we have simple conformity.

Furthermore, we have also here the link between understanding and applicationonce you understand why, and causal connections in general, you know which variables influence which. Then, you or your partner in the application program, can search the way to act upon the relevant variables to achieve the right effect. Application then verifies understanding and you have a virtuous circle

Finally, there is the issue of transmission of knowledge and the division of roles within it: teacher to pupil, coach to player, usually with pupils more passive but expected to be more active. The process of schooling typically combines all the features relevant to the transmission. All this prompts a very, very different picture of curiosity and of virtues than the one dominating the individual epistemology.

Naturalism

The third among the "further directions" for the inquiry has to do with naturalization.

The naturalization project has been present in the twentieth-century epistemology under various guises. Indeed, naturalizing epistemic virtue is a project that has until now produced a couple of books. When talking about the relation of curiosity to cognitive psychology, we briefly mentioned the prospect of naturalizing curiosity, which will interest us here. The context is the wider project of the dialogue between philosophy and science, in particular cognitive psychology, and issue of naturalization of philosophical insights. In the present case we can note that cognitive science supports an important part of our picture of curiosity. Let us look at it briefly.

The very notion of naturalization has several distinct aspects. A general one has to do with relation to science: to naturalize an activity, in our example, the cognitive effort, is to relate it to scientific investigation. Here is, for example, Abrol Fairweather talking about epistemic virtues:

> Bridges Between Virtue Epistemology and Philosophy of Science
>
> Bridge 1: Empirically Informed Theories of Epistemic Virtue
>
> One clear way of bridging work on epistemic virtue and work in the relevant sciences is to grounds their epistemic psychology (person level cognitive dispositions and the processes that count as their manifestation) in current work in evolutionary biology and cognitive science. (Fairweather ed., 2014: 3)

The vast literature on naturalization has produced some interesting proposals, which we can use for our curiosity topic. Following Klemens Kappel (2011) we shall assume that the important distinction is the one between descriptive-explanatory (Kappel calls it just "descriptive") and normative naturalization. The descriptive-explanatory task is to find scientifically acceptable grounding for philosophical characterizations: in our case we have analyzed the functioning of an epistemic virtue, curiosity, and can now ask whether cognitive psychology (together with evolutionary theory) would accept the basic characterization we have been working with, and how it would situate it in its framework, in particular its explanatory component. We just quoted a relevant idea from Abrol Fairweather, to the effect that naturalizing epistemic virtues means "grounding their epistemic psychology (person level cognitive dispositions and the processes that count as their manifestation) in current work in evolutionary biology and cognitive science". I hope that our Chap. 8 on cognitive science of curiosity documents that fact that there are no big problems in the cognitive account of the matters.

The more challenging kind is normative naturalization. Kappel offers a constructive proposal, inspired, I think, by Goldman's reliabilism:

> Consider then the problem of evaluating our epistemic norms. Assume that according to reliabilism, an epistemic norm is proper just if reliable under the conditions of its normal use. So certifying that some epistemic norm is proper requires canvassing empirical evidence of its reliability under those circumstances. Explaining the features in virtue of which the norm is proper would be accomplished by pointing to the features that make it reliable. Say that some subject S is justified in relying on an epistemic norm just in case doing so results from a reliable process for the selection of proper epistemic norms to rely upon. (2011: 843)

In his 2014 paper David Copp similarly points to the fact that we need epistemic norms in order to solve epistemic problems we face, and stresses the plurality of the possibly relevant systems of norms (2014: 74 ff.). We shall be following the methodology sketched by Kappel (2011) and Copp (2014).

Well, imagine we found that curiosity is epistemically successful, and that its success is due to its specific structure, revealed also by cognitive research on curiosity. Then, we would be able to say that norms leading to this success are the right means to the valuable end. This would then offer a beginning of a normative naturalization of the activities of human curiosity. Here is the application of the Copp-Kappel line to understanding-directed curiosity:

> Assume that a norm is proper just if successful under the conditions of its normal use.
>
> Assume also that desire to understand some item I is directed to finding a model of I which makes clear the important causal, logical and metaphysical connections that explain "various kinds of dependency relations". (As Greco would put it in his 2014 chapter)

Now, assume optimistically that the desire to understand is the motivation to 'cope' with I in the sense just sketched. "I don't understand the accident that happened, I don't know why the house burned down" or, to use a more up-to-date example, "I don't understand why corona virus spread so quickly in Italy." Well, the epidemiologist can draw a model for you, stressing the temporal delay of medical reaction of Italian establishment, the dynamic of movement of Italian citizens, and so on. After a while, you will be able to 'cope' with the enigma of the spread of corona virus. The norm tells you to look for causes, medical, demographic and perhaps institutional.

Understanding some different target, say a problem in geometry, would involve finding a model of spatial-mathematical relation governing the domain. Again, the desire to understand is the desire to cope, that is, to find such a model. (The history of geometry, from pharaonic Egypt to Hellenistic epoch, is the history of the search of a model that is sufficiently general and reliable. The reliable success offers also a naturalistic teleological account of the virtuous epistemic process involved.)

To apply the paradigm we decided to follow, explaining the features in virtue of which the norm for understanding is proper would be accomplished by pointing to the features that make it reliably successful. As Kappel writes, "Say that some subject S is justified in relying on an epistemic norm just in case doing so results from a reliable process for the selection of proper epistemic norms to rely upon" (2011: 843).

But is this enough for naturalization?—a more radical naturalist might ask. Here is a possible answer. Remember that we assumed that the cognitive processes involved to have a known cognitive explanation; this is guaranteed by descriptive-explanatory naturalization. What is then left, but the explanation of the success of the norm(s) governing them? And this is certainly available in the case of understanding and of the desire to understand.

To summarize, the naturalization project has two aspects, the descriptive-explanatory and the normative one. The first corresponds to what Fairweather describes as the "first bridge" between virtue epistemology and philosophy of science (2014: 53), namely grounding the virtuous cognitive dispositions in "current work in evolutionary biology and cognitive science" (2014: 3). The normative line is weakly teleological, and the two together should be able to ground the naturalization of curiosity.

In short, we have been bringing together, in a rather sketchy way, for which we apologize, the work of some of the authors in the field; interestingly, nothing specific has been written on curiosity in this context. This is one important direction for further inquiry.

References

Baehr, Jason. 2011. *The Inquiring Mind: On Intellectual Virtues and Virtue, Epistemology*. Oxford: Oxford University Press.

Copp, David. 2014. Indirect Epistemic Teleology Explained and Defended. In *Naturalizing Epistemic Virtue*, ed. Abrol Fairweather and Owen Flanagan, 70–91. Cambridge University Press.

Elgin, Catherine Z. 1996. *Considered Judgment*. Princeton, NJ: Princeton University Press.

Fairweather, Abrol, ed. 2014. *Virtue Epistemology Naturalized Bridges between Virtue Epistemology and Philosophy of Science*. Springer.

Grimm, Stephen R. 2011. Understanding. In *The Routledge Companion to Epistemology*, ed. Sven Bernecker and Duncan Pritchard, 84–94. New York: Routledge.

Kappel, Klemens. 2011. Naturalistic Epistemology. In *The Routledge Companion to Epistemology*, ed. Sven Bernecker and Duncan Pritchard, 836–846. New York: Routledge.

Kvanvig, Jonathan L. 2003. *The Value of Knowledge and the Pursuit of Understanding*. Cambridge University Press.

———. 2012. Curiosity and a Response-Dependent Account of the Value of Understanding. In *Knowledge, Virtue, and Action: Putting Epistemic Virtues to Work*, ed. T. Henning and D. Schweikard, 151–164. London: Routledge.

Pritchard, Duncan. 2010. Knowledge and Understanding. In *The Nature and Value of Knowledge, Three Investigations*, ed. D. Pritchard, A. Millar, and A. Haddock. Oxford University Press.

Riggs, Wayne D. 2009. Understanding, Knowledge, and the Meno Requirement. In *Epistemic Value*, ed. Adrian Haddock, Alan Millar, and Duncan Pritchard. Oxford: Oxford University Press.

Sosa, Ernest. 2001. For the Love of Truth? In *Virtue Epistemology: Essays on Epistemic Virtue and Responsibility*, ed. Abrol Fairweather and Linda Zagzebski. Oxford University Press.

———. 2007. *A Virtue Epistemology Apt Belief and Reflective Knowledge, Volume I*. Clarendon Press.

References

Adair, Gene. 1996. *Thomas Alva Edison: Inventing the Electric Age*. Oxford University Press.

Ahbel-Rappe, Sara, and Rachana Kamtekar, eds. 2006. *Blackwell Companion to Socrates*. Blackwell.

Alicke, Marc D., and David A. Dunning, eds. 2005. *The Self in Social Judgment*. Psychology Press.

Alston, William. 2005. *Beyond Justification: Dimensions of Epistemic Evaluation*. Cornell University Press.

Annas, Julia. 1985. Self-knowledge in Early Plato. In *Platonic Investigations, Studies in Philosophy and the History of Philosophy*, vol. 13, ed. Dominic J. O'Meara, 111–138. Catholic University of America.

Annas, Julia, and Jonathan Barnes. 2000. *Sextus Empiricus: Outlines of Scepticism*, Cambridge Texts in the History of Philosophy. 2nd ed. Cambridge University Press.

Augustine. 2006. *Confessions and Enchiridion* (The Library of Christian Classics). Trans. and ed. A.C. Outler. Westminster John Knox Press. https://www.ccel.org/ccel/augustine/confessions.xiii.html#fnf_xiii-p177.1.

Baehr, Jason. 2011. *The Inquiring Mind: On Intellectual Virtues and Virtue, Epistemology*. Oxford: Oxford University Press.

Ball, Philip. 2012. *Curiosity: How Science Became Interested in Everything*. The University of Chicago Press.

N. Miščević, *Curiosity as an Epistemic Virtue*, Palgrave Innovations in Philosophy,
https://doi.org/10.1007/978-3-030-57103-0

Barnes, Jonathan. 1975. *Aristotle's Posterior Analytics*. Oxford University Press.

Battaly, Heather. 2014. Acquiring Epistemic Virtue Emotions, Situations, and Education. In *Naturalizing Epistemic Virtue*, ed. Abrol Fairweather and Owen Flanagan, 175–195. Cambridge University Press.

Baumgarten, Elias. 2001. Curiosity as a Moral Virtue. *International Journal of Applied Philosophy* 15: 169–184.

Berlyne, Daniel. 1954. A Theory of Human Curiosity. *British Journal of Psychology* 45 (3): 180–191.

Berlyne, Daniel E. 1960. *Conflict, Arousal and Curiosity*. McGraw Hill.

———. 1966. Curiosity and Exploration. *Science, New Series* 153 (3731): 25–33.

Blumenberg, Hans. 1966. *The Legitimacy of the Modern Age*. MIT Press.

———. 1988. *Der Prozeß der theoretischen Neugierde*. Suhrkamp.

Bradford, Gwen. 2015a. Knowledge, Achievement, and Manifestation. *Erkenntnis* 80: 97–116.

———. 2015b. *Achievement*. Oxford University Press.

Brady, Michael S. 2009. Curiosity and the Value of Truth. In *Epistemic Value*, ed. A. Millar Haddock and D. Pritchard, 265–284. Oxford University Press.

———. 2011. Interest and Epistemic Goodness. Handout.

Browne, Elizabeth Janet. 2002. *Darwin Power of Place*. Vol. 2. New York: Knopf.

Burnyeat, Miles F. 1980. Can the Sceptic Live His Scepticism? In *Doubt and Dogmatism*, ed. M. Schofield, M.F. Burnyeat, and J. Barnes, 20–53. Oxford: Clarendon Press; Reprinted in Burnyeat and Frede 1997: 25–57 (page references to reprint).

———. 1981. Aristotle on Understanding Knowledge. In *Aristotle on Science: 'The Posterior Analytics' (Proceedings of the Eighth Symposium Aristotelicum)*, E.B. Padua, ed., 97–139. Reprinted in 2012 *Explorations in Ancient and Modern Philosophy Volume* II, 115–143. Cambridge University Press.

———. 2011. Episteme. In *Episteme, etc.: Essays in Honour of Jonathan Barnes*, ed. B. Morison and K. Ierodiakonou. Oxford University Press.

Carter, J. Adam, Benjamin Jarvis, and Katherine Rubin. 2013. Knowledge: Value on the Cheap. *Australasian Journal of Philosophy* 91: 249–263.

Cassam, Quasim. 2014. *Self-Knowledge for Humans*. Oxford: Oxford University Press.

Cholbi, Michael. The Value of Self-Knowledge (web). https://cpp.academia.edu/MichaelCholbi.

Christianson, Gale E. 1996. *Isaac Newton and the Scientific Revolution*. Oxford University Press.

Cicero, M.T. 1913. *De Officiis*. With an English Translation. Walter Miller. Cambridge, MA: Harvard University Press.

Cicerón, Marco Tulio, M.T. Griffin, and E.M. Atkins. 1991. *Cicero: On Duties*, Cambridge Texts in the History of Political Thought. Cambridge: Cambridge University Press.

Cohen, Bernard I., and George E. Smith. 2004. *The Cambridge Companion to Newton*. Cambridge University Press.

Coliva, Analisa. 2012. *The Self and Self-Knowledge*. Oxford University Press.

———. 2016. *The Varieties of Self-Knowledge*. Palgrave Macmillan.

Copp, David. 2014. Indirect Epistemic Teleology Explained and Defended. In *Naturalizing Epistemic Virtue*, ed. Abrol Fairweather and Owen Flanagan, 70–91. Cambridge University Press.

Darwin, Charles. 1809. *Autobiography the Life and Letters of Charles Darwin—Volume I*. New York: W. W. Norton & Company.

———. 1848. Letter to John Stephens Henslow, April 1, Darwin Correspondence Project, Letter no. 1167. http://www.darwinproject.ac.uk/DCP-LETT-1167.

Daston, Lorraine. 2000. Preternatural Philosophy. In *Biographies of Scientific Objects*, ed. Lorraine Daston. University of Chicago Press.

———. 2001. *Eine kurze Geschichte der wissenschaftlichen Aufmerksamkeit*. München: C.F. von Siemens Stiftung.

———. 2011. The Empire of Observation, 1600–1800. In *Histories of Scientific Observation*, ed. L. Daston and E. Lunbeck. Chicago: The University of Chicago Press.

De Regt, Henk W. 2017. *Understanding Scientific Understanding*. Oxford: Oxford University Press.

De Regt, Henk W., Sabina Leonelli, and Kai Eigner. 2013. *Scientific Understanding: Philosophical Perspectives*. University of Pittsburgh Press.

Dunning, David. 2005. *Self-Insight: Roadblocks and Detours on the Path to Knowing Thyself*. New York: Psychology Press.

———. 2011. The Dunning–Kruger Effect: On Being Ignorant of One's Own Ignorance. In *Advances in Experimental Social Psychology*, vol. 44. Amsterdam: Elsevier Inc.

Einstein, Albert. 1955. Old Man's Advice to Youth: 'Never Lose a Holy Curiosity.' *LIFE Magazine*, 2 May. From the memoirs of William Miller, an editor, quoted in Life Magazine.

Elgin, Catherine Z. 1996. *Considered Judgment*. Princeton, NJ: Princeton University Press.

———. 2006. From Knowledge to Understanding. In *Epistemology Futures*, ed. S. Hetherington, 199–215. Oxford University Press.

———. 2009. Is Understanding Factive? In *Epistemic Value*, ed. A. Millar Haddock and D. Pritchard, 322–330. Oxford: Oxford University Press.

———. 2017. Exemplification in Understanding. In *Explaining Understanding New Perspectives from Epistemology and Philosophy of Science*, ed. Stephen R. Grimm, Christoph Baumberger, and Sabine Ammon. Routledge.

Engel, Susan L. 2015. *The Hungry Mind: The Origins of Curiosity in Childhood*. OUP.

Eurich, Tasha. 2017. *Insight: Why We're Not as Self-Aware as We Think, and How Seeing Ourselves Clearly Helps Us Succeed at Work and in Life*. Crown Business.

Evans, Gareth. 1982. *The Varieties of Reference*. Oxford: Oxford University Press.

Fadiman, Clifton, and André Bernard. 2000. *Bartlett's Book of Anecdotes*. Boston, MA: Little, Brown and Company; Revised, Subsequent Edition (September 1, 2000).

Fairweather, Abrol, ed. 2014. *Virtue Epistemology Naturalized Bridges between Virtue Epistemology and Philosophy of Science*. Springer.

Fairweather, Abrol, and Montemayor Carlos. 2018. Curiosity and Epistemic Achievement. In *The Moral Psychology of Curiosity*, ed. Lani Watson et al., 199–216. Rowman & Littlefield International.

Fairweather, Abrol, and Owen Flanagan, eds. 2014. *Naturalizing Epistemic Virtue*. Cambridge University Press.

Fellner, Gene, Wesley Pitts, and Mark Zuss. 2012. Beyond the Sensible World: A Discussion of Mark Zuss. The Practice of Theoretical Curiosity. *Cultural Studies of Science Education* 7: 745–776.

Fernández, Jordi. 2013. *Transparent Minds: A Study of Self-Knowledge*. Oxford: Oxford University Press.

Ferrari, Michael, and Robert J. Sternberg, eds. 1998. *Self-Awareness, Its Nature and Development*. New York and London: Guilford Press.

Firestein, Stuart. 2012. *Ignorance: How It Drives Science*. Oxford: Oxford University Press.

Fürst, Martina. submitted. *Implicit Bias and Qualiefs*.

Gana, Kamel, ed. 2012. *Psychology of Self-Concept*. New York: Nova Science Publishers.

Gibbons, John. 2013. *The Norm of Belief*. Oxford University Press.

Godfrey-Smith, P. 2003. *Theory and Reality*. Chicago: The University of Chicago Press.

Goldman, Alvin. 1978. Epistemics: The Regulative Theory of Cognition. *Journal of Philosophy* 75: 509–523.

———. 1979. What Is Justified Belief? In *Justification and Knowledge*, ed. G. Papas, 1–23. Dordrecht: D. Reidel.

———. 1986. *Epistemology and Cognition*. Cambridge, MA: Harvard University Press.

———. 1992a. Epistemic Folkways and Scientific Epistemology. In *Liaisons: Philosophy Meets the Cognitive and Social Sciences*, ed. A. Goldman. Cambridge, MA: MIT Press.

———. 1992b. *Liaisons: Philosophy Meets the Cognitive and Social Sciences*. Cambridge, MA: MIT Press.

———. 1994. Naturalistic Epistemology and Reliabilism. In *Midwest Studies in Philosophy. Philosophical Naturalism*, ed. P.A. French, J. Uehling, E. Theodore, and H.K. Wettstein. Notre Dame, IN: University of Notre Dame Press.

———. 2002. The Unity of the Epistemic Virtues. In *Pathways to Knowledge: Private and Public*, ed. A. Goldman. Oxford University Press.

———. 2005. Kornblith's Naturalistic Epistemology. *Philosophy and Phenomenological Research* 71: 403–410.

Goldman, Alvin I., and Moshe Shaked. 1991. An Economic Model of Scientific Activity and Truth Acquisition. *Philosophical Studies* 63: 31–55.

Gopnik, Allison. 2000. Explanation as Orgasm and the Drive for Causal Knowledge: The Function, Evolution, and Phenomenology of the Theory Formation System. In *Explanation and Cognition*, ed. Frank C. Keil and R.A. Wilson, 299–324. Cambridge, MA: MIT Press.

Graver, Margaret R. 2014. Honeybee Reading and Self-Scripting: *Epistulae Morales* 84. In *Seneca Philosophus*, ed. Jula Wildberger and Marcia L. Colish, 269–294. Walter de Gruyter.

Grimm, Stephen R. 2008. Epistemic Goals and Epistemic Values. *Philosophy and Phenomenological Research* 77 (3): 725–744.

———. 2011. Understanding. In *The Routledge Companion to Epistemology*, ed. Sven Bernecker and Duncan Pritchard, 84–94. New York: Routledge.

———. 2017. Understanding and Transparency. In *Explaining Understanding New Perspectives from Epistemology and Philosophy of Science*, ed. Stephen R. Grimm, Christoph Baumberger, and Sabine Ammon, 212–229. Routledge.

Haas, Jens, and Katja M. Vogt. 2015. Ignorance and Investigation. In *Routledge International Handbook of Ignorance Studies*, ed. M. Gross and L. McGoey, 17–25. London: Routledge.

Habermas, Jǘrgen. 1971. *Knowledge and Human Interests*. Boston: Beacon Press.

Haddock, Adrian, Alan Millar, and Duncan Pritchard, eds. *Epistemic Value*. Oxford: Oxford University Press.

Hargittai, Istvan. 2011. *Drive and Curiosity: What Fuels the Passion for Science*. Amherst: Prometheus Books.

Hatzimoysis, Anthony. 2011. Introduction. In *Self-Knowledge*, ed. A. Hatzimoysis. Oxford University Press.

Hazlett, A. 2013. *A Luxury of the Understanding: On the Value of True Belief*. Oxford University Press.

Hoyle, R.H. 2010. Personality and Self-Regulation. In *Handbook of Personality and Self-Regulation*, ed. R.H. Hoyle. Oxford: Blackwell.

Huff, Toby E. 2011. *Intellectual Curiosity and the Scientific Revolution: A Global Perspective*. Cambridge University Press.

Hurka, Thomas. 2001. *Virtue, Vice, and Value*. Oxford University Press.

Hursthouse, Rosalind. 2000. *On Virtue Ethics*. Oxford University Press.

Inan, Ilhan. 2012. *The Philosophy of Curiosity*. Routledge.

———. 2014. Curiosity, Belief and Acquaintance. In *Virtue Epistemology Naturalized: Bridges between Virtue Epistemology and Philosophy of Science*, ed. A. Fairweather, 143–158. Springer.

———. 2016. Curiosity and Ignorance. *Croatian Journal of Philosophy* XVI (48): 285–304.

Inan, Ilhan, Lani Watson, Dennis Whitcomb, and Safiye Yiğit, eds. 2018. *The Moral Psychology of Curiosity*. Rowman & Littlefield International Ltd.

Jones, Matthew L. 2006. *The Good Life in the Scientific Revolution Descartes, Pascal, Leibniz, and the Cultivation of Virtue*. Chicago: The University of Chicago Press.

Kamtekar, Ramchandra. 2017. Self-Knowledge in Plato. In *Self-Knowledge: A History*, Oxford Philosophical Concepts Series, ed. Ursula Renz, 25–43. Christia Mercer.

Kappel, Klemens. 2011. Naturalistic Epistemology. In *The Routledge Companion to Epistemology*, ed. Sven Bernecker and Duncan Pritchard, 836–846. New York: Routledge.

Kashdan, Todd B., and Paul J. Silvia. 2009. Curiosity and Interest: The Benefits of Thriving on Novelty and Challenge. In *Handbook of Positive Psychology*, ed. R. Snyder Charles and Shane J. Lopez, 367–374. Oxford University Press.

Kenny, Neil. 2004. *The Uses of Curiosity in Early Modern France and Germany*. Oxford University Press.

Kim, Jaegwon. 1988. What Is "Naturalized Epistemology"? In *Philosophical Perspectives, Epistemology*, ed. J.E. Tomberline. Atascadero, CA: Ridgeview Publishing Co.

Kim, Dong-Won. 2002. *Leadership and Creativity a History of the Cavendish Laboratory, 1871–1919*. Springer.

Kitcher, Peter. 1990. The Division of Cognitive Labor. *Journal of Philosophy* 87 (1): 5–22.

Kitcher, P. 1993. *The Advancement of Science*. Oxford: Oxford University Press.

Knorr, Cetina K. 1999. *Epistemic Cultures How the Sciences Make Knowledge*. Cambridge, MA: Harvard University Press.

Kounios, John, and Mark Beeman. 2015. *The Eureka Factor: Aha Moments, Creative Insight, and the Brain*. Random House.

Kraut, Richard. 2006. The Examined Life. In *Blackwell Companion to Socrates*, ed. Sara Ahbel-Rappe and Rachana Kamtekar. Oxford: Blackwell.

Kristjánsson, Kristján. 2018. *Virtuous Emotions*. Oxford University Press.

Kvanvig, Jonathan L. 2003. *The Value of Knowledge and the Pursuit of Understanding*. Cambridge University Press.

———. 2012. Curiosity and a Response-Dependent Account of the Value of Understanding. In *Knowledge, Virtue, and Action: Putting Epistemic Virtues to Work*, ed. T. Henning and D. Schweikard, 151–164. London: Routledge.

Lao Tzu, Tao. 2006. *Te Ching*: A New English Version (Perennial Classics) Paperback, September 5, 2006, translator Stephen Mitchel Harper Perennial Modern Classics; Reprint Edition.

Leigh, Matthew. 2013. *From Polypragmon to Curiosus: Ancient Concepts of Curious and Meddlesome Behaviour*. OUP.

Lipton, Peter. 2009. Understanding Without Explanation. In *Scientific Understanding: Philosophical Perspectives*, ed. H.W. De Regt, S. Leonelli, and K. Eigner, 43–63. Pittsburgh: University of Pittsburgh Press.

Lynch, Michael. 2004. *True to Life: Why Truth Matters*. MIT Press.

Manson, Neil C. 2012. Epistemic Restraint and the Vice of Curiosity. *Philosophy* 87 (340): 239–259.

McMullin, Ernan M. 1982. Values in Science. *PSA: Proceedings of the Biennial Meeting of the Philosophy of Science Association*. Volume Two: Symposia and Invited Papers, 3–28.

Merton, Robert K. 1957. Priorities in Scientific Discovery. In *The Sociology of Science Theoretical and Empirical Investigations*, 386–411. Chicago: The University of Chicago Press.

Miščević, Nenad. 2000. *Rationality and Cognition: Against Relativism-Pragmatism*. University of Toronto Press.

———. 2004. Response-Intentionalism about Color [Croatian]. *Journal of Philosophy* 4 (2): 179–191.

———. 2007. Virtue-based Epistemology and the Centrality of Truth, (Towards a Strong Virtue-Epistemology). *Acta Analytica* 22: 239–266.
———. 2016a. Curiosity—The Basic Epistemic Virtue. In *Moral and Intellectual Virtues in Western and Chinese Philosophy, The Turn toward Virtue*, ed. Chienkuo Mi, M. Slote, and E. Sosa, 145–163. Routledge.
———. 2016b. Epistemic Value-Curiosity, Knowledge and Response-Dependence. *Croatian Journal of Philosophy* 16 (3): 393–417.
———. 2016. Epistemic Value-Curiosity, Knowledge and Response-Dependence. *Croatian Journal of Philosophy* 15 (48): 393–417.
———. 2017. The Value of Self-Knowledge. In *Perspectives on the Self*, ed. B. Berčić. Rijeka: University of Rijeka.
———. 2018a. Curiosity, Its Objects and Varieties. In *The Moral Psychology of Curiosity*, ed. I. Inan, L. Watson, D. Whitcomb, and S. Yiğit, 35–55. London: Rowman & Littlefield.
———. 2018b. Curiosity, Its Objects and Varieties. In *Moral Psychology of Curiosity*, ed. Lany Watson. Rowman and Littlefield.
———. 2018c. Self-inquisitiveness. *Acta Analytica* 33 (3): 331–352.
Moore, Christopher. 2015. *Socrates and Self-Knowledge*. Cambridge, UK: Cambridge University Press.
Muldoon, Ryan. 2017–2018. Diversity, Rationality, and the Division of Cognitive Labor. In *Scientific Collaboration and Collective Knowledge*, ed. T.H. Boyer-Kassem, C. Mayo-Wilson, and M. Weisberg, 78–93. Oxford: Oxford University Press.
Natsoulas, Thomas. 2015. *The Conceptual Representation of Consciousness*. Cambridge, UK: Cambridge University Press.
Newton-Smith, William. 2001. *Companion to Philosophy of Science*. Oxford: Blackwell.
Nozick, Robert. 1989. *Examined Life*. Simon and Shuster.
Nussbaum, Martha Craven. 1947 [1986]. *The Fragility of Goodness: Luck and Ethics in Greek Tragedy and Philosophy*. Cambridge University Press.
———. 1998. *Cultivating Humanity*. Liberal Education, Series.
Olfert, Christiana M.M. 2015. Skeptical Investigation and Its Perks: Diog. Laert. 9.69–70 and 79–89. In *Pyrrhonian Skepticism in Diogenes Laertius*, ed. Vogt Katja Maria. Tübingen: Mohr Siebeck.
Pais, Abraham. 1991. *Niels Bohr*. Oxford: Clarendon Press.
Papastephanou, Marianna, ed. 2019. *New Philosophical Explorations of the Epistemic Desire to Know: Just Curious About Curiosity*. Cambridge Scholars Publishing.

Patrick, Sean. 2013. *Nikola Tesla: Imagination and the Man That Invented the 20th Century*. Clearwater: Oculus Publishers.

Pennock, Robert T. 2019. *An Instinct for Truth Curiosity and the Moral Character of Science*. MIT Press.

Peterson, Christopher, and Martin E. Seligman. 2004. *Character Strengths and Virtues a Handbook and Classification*. OUP.

Pritchard, Duncan. 2010. Knowledge and Understanding. In *The Nature and Value of Knowledge, Three Investigations*, ed. D. Pritchard, A. Millar, and A. Haddock. Oxford University Press.

———. 2011. What is the Swamping Problem? In *Reasons for Belief*, ed. Andrew Reisner and Asbjørn Steglich-Petersen, 244–259. Cambridge University Press.

———. 2014. Truth as the Fundamental Epistemic Good. In *The Ethics of Belief: Individual and Social*, ed. J. Matheson and R. Vitz. Oxford University Press.

———. 2018. Epistemic Axiology. In *Epistemic Reasons, Epistemic Norms, and Epistemic Goals*, ed. M. Grajner and P. Schmechtig. DeGruyter.

Pritchard, D. 2016. Epistemic Axiology. In *Epistemic Reasons, Epistemic Norms, and Epistemic Goals*, ed. M. Grajner and P. Schmechtig. Berlin: DeGruyter.

Rechenberg, Helmut. 2010. *Werner Heisenberg–Die Sprache derAtome Leben und Wirken—Eine wissenschaftliche Biographie Die "FröhlicheWissenschaft" (Jugend bis Nobelpreis)*. Springer.

Renz, Ursula. 2017. Introduction. In *Self-Knowledge: A History*, ed. U. Renz. Oxford: Oxford University Press.

Riggs, Wayne D. 2009. Understanding, Knowledge, and the Meno Requirement. In *Epistemic Value*, ed. Adrian Haddock, Alan Millar, and Duncan Pritchard. Oxford: Oxford University Press.

Roberts, Robert C., and W. Jay Wood. 2007. *Intellectual Virtues*. Oxford University Press.

Rowe, Christopher. 2011. Self-Examination. In *The Cambridge Companion to Socrates*, ed. Donald R. Morrison, 201–214. Cambridge University Press.

Schmitt, Frederick F., and Reza Lahroodi. 2008. The Epistemic Value of Curiosity. *Educational Theory* 58 (2): 125–148.

Shapin, Steven. 1996. *Scientific Revolution*. Chicago: University of Chicago Press.

Shapin, Steven, and Simon Schaffer. 1985. *Leviathan and the Air-Pump Hobbes, Boyle, and the Experimental Life*. Princeton University Press.

Silvia, Paul J. 2006. *Exploring the Psychology of Interest*. Oxford University Press.

———. 2008. Interest—The Curious Emotion. *Current Directions in Psychological Science* 17 (1): 57–60.

———. 2012. Curiosity and Motivation. In *The Oxford Handbook of Human Motivation*, ed. M. Ryan Richard, 1st ed. Oxford University Press.

Sosa, Ernest. 2001. For the Love of Truth? In *Virtue Epistemology: Essays on Epistemic Virtue and Responsibility*, ed. Abrol Fairweather and Linda Zagzebski. Oxford University Press.

———. 2007. *A Virtue Epistemology Apt Belief and Reflective Knowledge, Volume I*. Clarendon Press.

———. 2015. *Judgment and Agency*. Oxford University Press.

Stalnaker, Robert C. 2008. *Our Knowledge of the Internal World*. Oxford: Clarendon Press.

Stanley, Jason, and Timothy Williamson. 2001. Knowing How. *The Journal of Philosophy* 98 (8): 411–444.

Stich, Stephen. 1990. *The Fragmentation of Reason: Preface to a Pragmatic Theory of Cognitive Evaluation*. MIT Press.

Stratton-Lake, Philip. 2002. Introduction. In *The Right & the Good*, ed. David Ross. Oxford: Clarendon Press.

Strevens, Michael. 2003. The Role of the Priority Rule in Science. *Journal of Philosophy* 100: 55–79.

Suster, Danilo. 2019. *What are Philosophers Doing?* Zbirka znanstvena monografija. Kulturni Center Maribor.

Tappenden, Jeremy. 2005. Proof Style and Understanding. In *Visualization, Explanation and Reasoning Style in Mathematics*, ed. Paul Mancosu, K.F. Jørgensen, and S.A. Pedersen, 147–210. Springer.

Thorsrud, Harald. 2012. Radical and Mitigated Skepticism in Cicero's *Academica*. In *Cicero's Practical Philosophy*, ed. Walter Nicgorski, 133–151. University of Notre Dame Press.

Ulrich Wiesmann, et al. 2013. Validating the Grand Five of the Generalized Health-Related Self-Concept. In *Psychology of Self-Concept 2012*, ed. Kamel Gana, 7–20. New York: Nova Science Publishers.

Veroff, Joseph, and Joanne B. Veroff. 2016. *Social Incentives: A Life-Span Developmental Approach*. Academic Press.

Walsh, Peter. 1988. The Rights and Wrongs of Curiosity (Plutarch to Augustine). *Greece & Rome* 35 (1): 73–85.

Weatherall, Andrew K.M. An Interview with Jocelyn Bell-Burnell (web). http://weatheralltech.com/bell/index.html.

Whitcomb, Dennis. 2018. Some Epistemic Roles for Curiosity. In *The Moral Psychology of Curiosity*, ed. Ilhan Inan, Lani Watson, Dennis Whitcomb, and Safiye Yiğit, 217–237. Rowman & Littlefield International Ltd.

Wildberger, Jula. 2006. Seneca and the Stoic Theory of Cognition: Some Preliminary Remarks. In *Seeing Seneca Whole Perspectives on Philosophy, Poetry and Politics*, ed. Katharina Volk and Gareth D. Williams, 75–102. Brill.

Williams, Michael. 2010. Descartes' Transformation of the Sceptical Tradition. In *The Cambridge Companion to Ancient Scepticism*, ed. Richard Arnot Home Bett. Cambridge University Press.

Winnicott, Donald W. 1965. *The Maturational Processes and the Facilitating Environment*. International Universities Press.

Wojcik, Jerzy W. 2000. Pursuing Knowledge. In *Rethinking the Scientific Revolution*, ed. Margaret J. Osier. Cambridge University Press.

Woolf, Raphael. 2012. *Cicero: The Philosophy of a Roman sceptic*. Routledge.

Wootton, David. 2015. *The Invention of Science: A New History of the Scientific Revolution*. New York: Harper.

Yiğit, Safiye. 2018. Curiosity as an Intellectual Virtue. In *The Moral Psychology of Curiosity*, ed. Ilhan Inan, Lani Watson, Dennis Whitcomb, and Safiye Yiğit, 117–140. Rowman & Littlefield International Ltd.

Yigit, Safiye. 2019. The Curious Case of Curiosity: A Virtue or a Vice? In *Toward New Philosophical Explorations of the Epistemic Desire to Know: Just Curious about Curiosity*, ed. Marianna Papastephanou, 150–165. Cambridge Scholars Publishing.

Zagzebski, Linda. 2001. Recovering Understanding. In *Knowledge, Truth, and Obligation: Essays on Epistemic Justification, Virtue, and Responsibility*, ed. M. Steup. Oxford: Oxford University Press.

———. 2003. Intellectual Motivation and the Good of Truth. In *Intellectual Virtue: Perspectives from Ethics and Epistemology*, ed. Michael DePaul and Linda Zagzebski, 135–154. Oxford University Press.

Zollman Kevin, J.S. 2017. Learning to Collaborate. In *Scientific Collaboration and Collective Knowledge*, ed. Thomas H. Boyer-Kassem, Conor Mayo-Wilson, and Michael Weisberg. Oxford University Press.

Zurn, Peter. 2019. Busybody, Hunter, Dancer: Three Historico-Philosophical Models of Curiosity. In *Curious about Curiosity: Toward New Philosophical Explorations of the Desire to Know*, ed. M. Papastefanou, 26–49. Cambridge Scholars Press.

Zuss, Mark. 2012. *The Practice of Theoretical Curiosity*. Springer.

Index

N. Miščević, *Curiosity as an Epistemic Virtue*, Palgrave Innovations in Philosophy,
https://doi.org/10.1007/978-3-030-57103-0

The manufacturer's authorised representative in the EU is Springer Nature Customer Service Centre GmbH, Europaplatz 3, 69115 Heidelberg, Germany. If you have any concerns regarding our products, please contact ProductSafety@springernature.com

Printed and bound by CPI Group (UK) Ltd, Croydon, CR0 4YY
22/09/2026
02221054-0009